BRITISH RAILW

LOCO

CW00344997

FORTIETH EDITION
SPRING 1998

The Complete Guide to all
Locomotives which run on
Britain's Mainline Railways and
Locomotives of Eurotunnel

Peter Fox and Richard Bolsover

ISBN 1 872524 24 9

CONTENTS

ACQUISITION OF INFORMATION

This book has been published with great difficulty. Privatisation of the railways and the splitting up of BR into different companies has been used as an excuse to deny the railway press access to official rolling stock library information, breaking a tradition of freely-supplied information which has existed for around half a century. We hope that readers will find the information accurate, but cannot be responsible for any inaccuracies.

We would like to thank the companies and individuals which have been co-operative in supplying information and would ask other companies which find this book useful to help us in future to make the book as accurate as possible.

This book is updated to 1st January 1998.

ORGANISATION OF BRITAIN'S RAILWAY SYSTEM

INFRASTRUCTURE

Britains national railway infrastructure, i.e. the track, signalling, stations and overhead line equipment is now owned by a private company called 'Railtrack'. Many stations and maintenance depots are leased to train operating companies. The exception is the infrastructure on the Isle of Wight, which is leased from the Government to Island Line.

DOMESTIC PASSENGER TRAIN OPERATIONS

Passenger trains are operated by train operating companies (TOCs). These TOCs operate on fixed term franchises. A list of these is appended below:

TOC	Operator	New Name
Anglia Railways	GB Trains	
Inter City East Coast	Sea Containers Ltd.	Great North Eastern Railway
Inter City West Coast	Virgin Group	Virgin Trains
Cross Country Trains	Virgin Group	Virgin Trains
Great Western Trains	Great Western Holdings	
North West Regional Railways	Great Western Holdings	North Western Trains
Midland Main Line	National Express	
Gatwick Express	National Express	
North London Railways	National Express	Silverlink
Central Trains	National Express	
Scotrail	National Express	
Merseyrail Electrics	MTL Holdings	
Regional Railways North East	MTL Holdings	
LTS Rail	Prism Rail	
South Wales & West Railway	Prism Rail	Wales & West Passenger Trains
Cardiff Railway Co.	Prism Rail	
West Anglia Great Northern	Prism Rail	
South West Trains	Stagecoach	
Island Line	Stagecoach	
Network South Central	Connex	Connex South Central
South East Trains	Connex	Connex South Eastern
Great Eastern	FirstBus	
Thameslink	GOVIA	
Chiltern Railways	M40 Trains	
Thames Trains	Victory Rail	

NOTES ON TRAIN OPERATING COMPANY OWNERS

Connex

This is a French company owned by Société Générale des Entreprises Automobiles, a subsidiary of Compagnie Générale des Eaux.

FirstBus

This is a large bus company which was originally formed by the amalgamation of Badgerline and GRT bus group.

GB Trains

This is a company set up for rail privatisation.

GOVIA

A joint venture between the Go-Ahead bus company and VIA, a French public transport operating company.

Great Western Holdings

This is a jointly owned by the former Great Western Trains management. 3i plc and FirstBus.

National Express

This is a transport operator which runs express coach services by sub-contracting them to various bus companies. It also owns east Midlands Airport.

M40 Trains

This is owned by the former management of Chiltern Railways.

MTL Holdings

This is the former municipal bus operator Merseyside PTE which operates buses in Merseyside and London.

Prism

This is a company whose shares are owned by individuals and financial institutions. Its chairman is joint managing director of EYMS, a bus company.

Sea Containers

This is a Bermuda-based shipping company which also owns the Venice-Simplon-Orient Express.

Stagecoach

Tha largest private bus operator in the UK.

Victory Railway Holdings

This is a joint venture between the Go Ahead group and Thames Trains managament.

Virgin Group

This is the well-known company headed by Richard Branson which has interests in travel, leisure and retailing.

CHANNEL TUNNEL PASSENGER TRAIN OPERATIONS

Eurostar trains are operated by Eurostar (UK) Ltd. jointly with French Railways (SNCF) and Belgian Railways (NMBS/SNCB). Eurostar (UK) will also operate the Night Service trains jointly with SNCF, Netherlands Railways (NS) and German Railways (DB).

FREIGHT TRAIN OPERATIONS

The three trainload freight companies Loadhaul, Mainline and Transrail which were set up on the goverment's orders in readiness for privatisation and Railfreight Distribution have been sold to the North & South Railway Company whose main shareholder is Wisconsin Central Transportation Corporation of the USA. Rail Express Systems, which operates mail and charter trains has also been sold to this company. The four concerns have been combined and now known as the English. Welsh and Scottish Railway Ltd. (EWS).

The container train operation known as Freightliner has been sold to a managment buyout known as Freightliner (1995) Ltd.

Certain other companies e.g. Direct Rail Services and National Power operate freight trains with their own locomotives.

OWNERSHIP OF LOCOMOTIVES AND ROLLING STOCK

The locomotives of EWS and those of Eurostar are owned by those companies. Most locomotives, hauled coaching stock and multiple unit vehicles used by the passenger train operating companies are owned by three leasing companies which were originally set up by British Railways as subsidiaries and then privatised. These are:

- Forward Trust Rail (formerly Eversholt Leasing) owned by Hong Kong Shanghai Bank.
- Angel Train Contracts (owned by the Royal Bank of Scotland).
- Porterbrook Leasing Company Ltd (owned by Stagecoach Holdings).

Other vehicles are owned or operated on behalf of private owners by various private companies such as Fragonset Railways, Carnforth Railway Restoration & Engineering Services Ltd., Titanstar Ltd. and the Venice-Simplon-Orient Express Ltd.

Further details of these companies will be found in the section on abbreviations and codes. Thus for each vehicle it is generally necessary to specify both the owner and the TOC which currently operates the vehicle.

A number of 'service' type vehicles are owned by Railtrack (e.g. Sandite vehicles) and others are owned by former BR Headquarters organisations which have now been privatised e.g. Railtest or by railway vehicle manufacturing and repair companies. Royal Train vehicles are owned by Railtrack.

INTRODUCTION

The following notes are applicable to locomotives:

DETAILS & DIMENSIONS

Principal details and dimensions are given for each class in metric units. Imperial equivalents are also given for power. Maximum speeds are still quoted in miles per hour since imperial units are still used in day to day railway operations in Britain.. Since the present maximum permissible speed of certain classes of locomotives is different from the design speed, these are now shown separately in class details. In some cases certain low speed limits are arbitrary and may occasionally be raised when necessary if a locomotive has to be pressed into passenger service.

LOCOMOTIVE DETAIL DIFFERENCES

Detail differences which affect the areas and types of train which locos work are shown. Where detail differences occur within a class or part class of locomotives., these are shown against the individual locomotive number. Except where shown, diesel locomotives have no train heating equipment and have train air brakes only. Electric or electro-diesel locomotives are assumed to have train heating unless shown otherwise. Standard abbreviations used are:

c	Fitted with Scharfenberg couplers for Eurostar working.
e	Fitted with electric heating apparatus (ETH).
j	Fitted with RCH jumper cables for operating with PCVs (propelling control vehicles)
r	Fitted with radio electronic token block equipment.
s	Slow speed control fitted (and operable).
t	Fitted with automatic vehicle identification transponders.
v	Train vacuum brakes only.
x	Dual train brakes (air & vacuum).
y	ETH equipped but equipment isolated.
+	Extended range locos with Additional fuel tank capacity compared with others in class.

After the locomotive number are shown any notes regarding braking, heating etc., the livery code (in bold type), the pool code where applicable, the depot code and name if any. Locomotives which have been renumbered in recent years show the last number in parentheses after the current number. For previous numbers of other locos, please refer to the Platform 5 Book 'Diesel & Electric Loco Register'.

NAMES

All official names are shown as they appear on the locomotive i.e. all upper case or upper & lower case lettering. Where only a few locomotives in a class are named, these are shown in a separate table at the end of the class or sub-class.

DEPOT ALLOCATIONS

The depot at which a locomotive is allocated is the one at which it receives its main examinations. This depot may be a long way away from where it normally performs its duties. Pool codes were allocated under the TOPS system and are still used, but have now become superfluous as, now that English Welsh & Scottish Railway have changed to a common user system for locos, locos of the same type at the same depot have the same pool code. A list of pool codes will be found on page 82. (S) denotes stored serviceable and (U) stored unserviceable. For stored locos the last known storage location is shown. Thus the layout is as follows:

No.	Old No.	Notes	Livery	Owner	Operation	Depot	Name
47777	(47636)	**RX**	E	EW	CD	Restored

GENERAL INFORMATION ON BRITISH RAILWAYS' LOCOMOTIVES

CLASSIFICATION & NUMBERING

Initially BR diesel locomotives were allocated numbers in the 1xxxx series, with electrics allotted numbers in the 2xxxx series. Around 1957 diesel locomotives were allocated new numbers with between one and four digits with 'D' prefixes. Diesel electric shunters in the 13xxx series had the '1' replaced by a 'D', but diesel mechanical shunters were completely renumbered. Electric locomotives retained their previous numbers but with an 'E' prefix.

When all standard gauge steam locomotives had been withdrawn, the prefix letter was removed. In 1972, the present TOPS numbering system was introduced whereby the loco number consisted of a two-digit class number followed by a serial number. In some cases the last two digits of the former number were generally retained (Classes 20, 37, 50), but in other classes this is not the case. In this book former TOPS numbers carried byrecently- converted locos. are shown in parentheses. Full renumbering information is to be found in the 'Diesel & Electric loco Register', the new third edition of which is now available.

Diesel locomotives are classified as 'types' depending on their engine horsepower as follows:

Type	Engine hp.	Old Number Range	Current Classes
1	800-1000	D 8000-D 8999	20.
2	1001-1499	D 5000-D 6499/D 7500-D 7999	31.
3	1500-1999	D 6500-D 7499	33, 37.
4	2000-2999	D 1-D 1999	46, 47, 50.
5	3000+	D 9000-D 9499	55, 56, 58, 59, 60.
Shunter	300-799	D 3000-D 4999	08, 09.

Class 14 (650 hp diesel hydraulics) were numbered in the D95xx series.

Electric and electro-diesel locomotives are classified according to their supply system. Locomotives operating on a d.c. system are allocated classes 71-80, whilst a.c. or dual voltage locomotives start at Class 81. Departmental locomotives which remain self propelled or which are likely to move around on a day to day basis are classified Class 97.

WHEEL ARRANGEMENT

For main line diesel and electric locomotives the system whereby the number of driven axles on a bogie or frame is denoted by a letter (A=1, B=2, C=3 etc.) and the number of undriven axles is noted by a number is used. The letter 'o' after a letter indicates that each axle is individually powered and a + sign indicates that the bogies are intercoupled. For shunters the Whyte notation is used. In this notation, generally used in Britain for steam locomotives, the number of leading wheels are given, followed by the number of driving wheels and then the trailing wheels.

HAULING CAPABILITY OF DIESEL LOCOS

The hauling capability of a diesel locomotive depends basically upon three factors:

1. Its adhesive weight. The greater the weight on its driving wheels, the greater the adhesion and thus more tractive power can be applied before wheel slip occurs.

2. The characteristics of its transmission. In order to start a train the locomotive has to exert a pull at standstill. A direct drive diesel engine cannot do this, hence the need for transmission. This may be mechanical, hydraulic or electric. The current British standard for locomotives is electric transmission. Here the diesel engine drives a generator or alternator and the current produced is fed to the traction motors. The force produced by each driven wheel depends on the current in its traction motor. In other words the larger the current, the harder it pulls.

As the locomotive speed increases, the current in the traction motors falls hence the *Maximum Tractive Effort* is the maximum force at its wheels that the locomotive can exert at a standstill. The electrical equipment cannot take such high currents for long without overheating. Hence the *Continuous Tractive Effort* is quoted which represents the current which the equipment can take continuously.

3. The power of its engine. Not all of this power reaches the rail as electrical machines are approximately 90% efficient. As the electrical energy passes through two such machines (the generator/alternator and the traction motors), the *Power At Rail* is about 81% (90% of 90%) of the engine power, less a further amount used for auxiliary equipment such as radiator fans, traction motor cooling fans, air compressors, battery charging, cab heating, ETH, etc. The power of the locomotive is proportional to the tractive effort times the speed. Hence when on full power there is a speed corresponding to the continuous tractive effort.

HAULING CAPABILITY OF ELECTRIC LOCOS

Unlike a diesel locomotive, an electric locomotive does not develop its power on board and its performance is determined only by two factors, namely its weight and the characteristics of its electrical equipment. Whereas a diesel locomotive tends to be a constant power machine, the power of an electric loco-motive varies considerably. Up to a certain speed it can produce virtually a constant tractive effort. Hence power rises with speed according to the formula given in section 3 above, until a maximum speed is reached at which tractive effort falls, such that the power also falls. Hence the power at the speed corresponding to the maximum tractive effort is lower than the maximum.

BRAKE FORCE

The brake force is a measure of the braking power of a locomotive. This is shown on the locomotive data panels so that railway staff can ensure that sufficient brake power is available on freight trains.

TRAIN HEATING AND POWER EQUIPMENT

The standard system in use in Britain for heating loco hauled trains is by means of electricity and is now known as ETS (Electric train supply). Locomotives which were equipped to provide steam heating have had this equipment removed or rendered inoperable (isolated). Electric heat is provided from the locomotive by means of a separate alternator on the loco, except in the case of Class 33 which have a d.c. generator. The ETH Index is a measure of the electrical power available for train heating. All electrically heated coaches have an ETH index and the total of these in a train must not exceed the ETH power of a locomotive.

ROUTE AVAILABILITY

This is a measure of a railway vehicle's axle load. The higher the axle load of a vehicle, the higher the RA number on a scale 1 to 10. Each route on BR has an RA number and in theory no vehicle with a higher RA number may travel on that route without special clearance. Exceptions are made, however.

MULTIPLE AND PUSH-PULL WORKING

Multiple working between diesel locomotives in Britain has usually been provided by means of an electro-pneumatic system, with special jumper cables connecting the locos. A coloured symbol is painted on the end of the locomotive to denote which system is in use. Class 47s nos. 47701-17 used a time-division multiplex (t.d.m.) system which utilised the existing RCH (an abbreviation for the former railway clearing house, a pre-nationalisation standards organisation) jumper cables for push-pull working. These had in the past only been used for train lighting control, and more recently for public address (pa) and driver-guard communication. A new standard t.d.m. system is now fitted to all a.c. electric locomotives and other vehicles, enabling them to work in both push-pull and multiple working modes.

COMMUNICATION

Virtually all main line locomotives are now fitted with cab to shore radio communication. Where locomotives are fitted with train communication this is stated in the class headings.

1. DIESEL LOCOMOTIVES

CLASS 08   BR SHUNTER   0-6-0

Built: 1953–62 by BR at Crewe, Darlington, Derby, Doncaster or Horwich Works.
Engine: English Electric 6KT of 298 kW (400 hp) at 680 rpm.
Main Generator: English Electric 801.
Traction Motors: Two English Electric 506.
Max. Tractive Effort: 156 kN (35000 lbf).
Cont. Tractive Effort: 49 kN (11100 lbf) at 8.8 m.p.h.
Power At Rail: 194 kW (260 hp).   **Length over Buffers:** 8.92 m.
Brake Force: 19 t.   **Wheel Diameter:** 1372 mm.
Design Speed: 20 m.p.h.   **Weight:** 50 t.
Max. Speed: 15 or 20* m.p.h.   **RA:** 5.

Non-standard liveries:
08077 & 08785 are RFS grey with blue and yellow bodyside stripes.
08296, 08602, 08846 & 08943 are grey and carry numbers 001, D 3769, D 4144 and 002 respectively.
08414 is **D** with RfD brandings and also carries its former number D 3529.
08460 is light grey with a dark grey roof, black cab doors and window surrounds and 'TLF South East' branding.
08500 is red lined out in black & white.
08519 is BR black.
08527 is light grey with a black roof, blue bodyside stripe and 'Ilford Level 5' branding.
08593 is Great Eastern blue lined out in red and also carries its former number D 3760.
08601 is London Midland & Scottish Railway black.
08629 is Royal purple.
08642 is London & South Western Railway black and also carries its former number D 3809.
08649 is grey with blue, red and white bodyside stripes and also carries its former number D 3816.
08689 is **D** with Railfreight general markings.
08715 is in experimental dayglo orange livery.
08721 is blue with a red & yellow stripe ('Red Star' livery).
08730 is BR black.
08743 & 09903 are Trafalgar blue.
08805 is LMS maroon and also carries its former number 3973.
08867 is BR black.
08879 is turquoise with full yellow ends, black cab doors, black numbers on a yellow background and RfD brandings.
08883 is Caledonian blue.
08907 is London & North Western Railway black.
08938 is grey and red.

08616 carries its former number D 3783.
08830 is on long-term lease to the East Somerset Railway and carries its former number D 3998.

n Waterproofed for working at Oxley Carriage Depot.
z Fitted with buckeye adaptor at nose end for HST depot shunting.

Formerly numbered in series 3000–4192.

Class 08/0. Standard Design.

					Depot	Location/Sub-depot
08077	a	0	FL	*FL*	CD	Millbrook FLT
08296	x	0	AD	*AD*	ZC	
08331	x	GN	RF		ZB (U)	
08388	a	F	E		IM (U)	
08389	a		E	*EW*	AN	Wembley
08393	a	D	E	*EW*	AN	Wembley
08397	a	F	E	*EW*	CD	
08401	a	D	E	*EW*	IM	
08402	a	D	E	*EW*	CD	
08405	a	D	E	*EW*	IM	
08410	a	D	GW	*GW*	PM	
08411	a		E	*EW*	ML	
08413	a	D	E		TI (U)	
08414	a*	0	E	*EW*	OC	
08417	a		SO		DY (U)	
08418	a	F	E	*EW*	DR	
08428	a		E		BS (U)	
08441	a		E	*EW*	TO	
08442	a	F	E	*EW*	KY	
08445	a		E		IM (U)	
08448	a		E		BS (U)	
08449	a		E		TO (U)	
08451	x		VT	*VW*	WN	
08454	x		VT	*VW*	WN	
08460	a	0	E	*EW*	CD	
08466	a	F0	E		IM (U)	
08472	a		GN	*GN*	EC	
08480	a	G	E	*EW*	EH	
08481	x		E	*EW*	CF	MG
08482	a	D	E	*EW*	AN	Wembley
08483	a	D	GW	*GW*	PM	
08484	a	D	RC	*RC*	ZN	
08485	a		E	*EW*	CD	
08489	a	F	E		WA (U)	
08492	a		E	*EW*	TO	
08493	a		E	*EW*	CF	MG
08495	x		E	*EW*	TO	
08499	a	F	E	*EW*	KY	
08500	x	0	E	*EW*	DR	
08506	a		E	*EW*	ML	
08509	a	F	E	*EW*	DR	
08510	a		E	*EW*	DR	
08511	a		E	*EW*	TO	
08512	a	F	E	*EW*	DR	

08514	a		E	EW	DR	
08516	a	D	E	EW	TO	
08517	a		E		SF (U)	
08519	a	0	E		BS (U)	
08523	x	ML	E	EW	OC	
08525	x	F	MM	ML	NL	
08526	x		E	EW	OC	
08527	x	0	AD	AD	ZI	
08528	x	D	E	EW	TO	
08529	x		E	EW	TO	
08530	x	D	FL	FL	CD	Tilbury FLT
08531	x	D	FL	FL	CD	Tilbury FLT
08534	x	D	E	EW	ML	
08535	x	D	E	EW	TI	SY
08536	x		MM		DY (U)	
08538	x	D	E	EW	TO	PB
08540	x	D	E		ZB (U)	
08541	x	D	E	EW	OC	
08542	x	F	E	EW	BS	
08543	x	D	E	EW	BS	
08561	x		E	EW	ML	
08567	x		E	EW	BS	
08568	x		RC	RC	ZH	
08569	x		E	EW	AN	
08571	xz		GN	GN	EC	
08573	x		AD	AD	ZI	
08575	x	BS	FL	FL	CD	Millbrook FLT
08576	x		E	EW	CF	BZ
08577	x		E	EW	TE	TY
08578	x	R	E	EW	OC	
08580	x		E	EW	TO	PB
08581	x	BS	E		ZB (U)	
08582	a	D	E	EW	TE	
08585	x		FL	FL	CD	Basford Hall Yard
08586	a	F	E		AY (U)	
08587	x		E	EW	DR	
08588	xz	BS	MM	ML	NL	
08593	x	0	E	EW	TO	PB
08594	x		E		TO (U)	
08597	x		E	EW	KY	
08599	x		E		SP (U)	
08601	x	0	E	EW	BS	
08602	x	0	AD	AD	ZD	
08605	x		E	EW	KY	
08607	x		E		TO (U)	
08610	x		E		BS (U)	
08611	x	V	VT	VW	LO	
08616	x	G	CT	CT	TS	
08617	x		VT	VW	WN	
08619	x		E		LO (U)	
08622	x		E		ML (U)	

08623	x		E	*EW*	BS	
08624	x		FL	*FL*	CD	Coatbridge FLT
08625	x		E		CF (U)	
08628	x		E	*EW*	BS	
08629	x	**0**	RC	*RC*	ZN	
08630	x		E	*EW*	ML	
08632	x		E	*EW*	IM	
08633	x	**RX**	E	*EW*	HT	
08635	x		E	*EW*	OC	
08641	xz	**D**	GW	*GW*	LA	PZ
08642	x*	**0**	FL	*FL*	CD	Felixstowe North FLT
08643	xz	**D**	GW	*GW*	PM	
08644	xz	**M**	GW		LA	
08645	xz	**D**	GW	*GW*	LA	
08646	x	**F**	E	*EW*	OC	
08648	x*	**D**	GW		LA	
08649	x	**0**	WT	*WT*	ZG	
08651	x	**D**	E	*EW*	CF	
08653	x*	**RD**	E	*EW*	AN	
08655	x*	**F**	E		AN (U)	
08661	a	**RD**	E		AN (U)	
08662	x		E	*EW*	KY	
08663	a	**D**	GW	*GW*	LA	
08664	x		E	*EW*	EH	
08665	x		E	*EW*	IM	
08670	a		E	*EW*	ML	
08675	x	**F**	E	*EW*	ML	
08676	x		E		ZB (U)	
08682	x		AD	*AD*	ZF	
08683	x		E		CF (U)	
08685	x		E	*EW*	ML	
08689	a	**0**	E		OC (U)	
08690	x		MM	*ML*	DY	
08691	x	**G**	FL	*FL*	CD	Trafford Park FLT
08692	x		AD		ZC (U)	
08693	x		E		ML (U)	
08694	x		E	*EW*	AN	Wembley
08695	x		E	*EW*	CD	
08696	a	**D**	VT	*VW*	WN	
08697	x		MM	*ML*	DY	
08698	a		E		ZB (U)	
08699	x	**D**	AD	*AD*	ZC	
08700	a		E		SF (U)	
08701	x	**RX**	E	*EW*	CD	
08702	x		E	*EW*	OC	
08703	a		E	*EW*	AN	
08706	x		E	*EW*	KY	
08709	x		E	*EW*	OC	
08711	x	**RX**	E	*EW*	OC	
08713	a		E		ZB (U)	
08714	x	**RX**	E	*EW*	TO	PB

08715	v	0	E		SF (U)	
08718	x		E		AY (U)	
08720	a	D	E	*EW*	ML	
08721	x	0	VT	*VW*	LO	
08723	x		E		TO (U)	
08724	x		GN		ZB (U)	
08730	x	0	RC	*RC*	ZH	
08731	x		E		ML (U)	
08734	x		E		CF (U)	
08735	x		E	*EW*	ML	FW
08737	x	RD	E		AN (U)	
08738	x	D	E	*EW*	CD	
08739	x		E		AN (U)	
08740	x	F	E		SF (U)	
08742	x	RX	E	*EW*	CD	
08743	x	I	I	*IC*	BH	
08745	x	RD	FL	*FL*	CD	Ipswich Upper Yard
08746	x	D	E		ZB (U)	
08750	x		E		SF (U)	
08751	x	RD	E		TI (U)	
08752	x	C	E	*EW*	TO	PB
08754	x		SR	*SR*	IS	
08756	x	D	E	*EW*	CF	
08757	x	RX	E	*EW*	HT	
08758	x		E		SF (U)	
08762	x		SR	*SR*	IS	
08765	xn	D	E	*EW*	BS	
08768	x		E	*EW*	ML	
08770	a	D	E	*EW*	CF	MG
08773	x		E		TO (U)	
08775	x		E	*EW*	OC	
08776	a	D	E	*EW*	KY	
08780	x		GW	*GW*	LE	
08782	a		E	*EW*	KY	
08783	x		E	*EW*	KY	
08784	x		E		AN (U)	
08785	x	0	FL		ZB (U)	
08786	a	D	E	*EW*	CF	BZ
08790	x		VT	*VW*	LO	
08792	x		E	*EW*	CF	EX
08795	x	M	GW	*GW*	LE	
08798	x		E	*EW*	CF	Tavistock Junction
08799	x		E	*EW*	AN	
08801	x		E	*EW*	CF	MG
08802	x	RX	E	*EW*	CD	
08804	x		E	*EW*	BK	
08805	x	0	CT	*CT*	TS	
08806	a	F	E	*EW*	TE	TY
08807	x	BS	E	*EW*	CD	
08810	a		AR	*AR*	NC	
08813	a	D	E	*EW*	TE	

08815	x		E		SP (U)	
08817	x	**BS**	E		SP (U)	
08819	x	**D**	E	*EW*	CF	
08822	x	**M**	GW	*GW*	LE	
08823	a		AD		ZF (U)	
08824	a	**F**	E	*EW*	IM	
08825	a		E	*EW*	AN	Wembley
08826	a		E		ML (U)	
08827	a		E	*EW*	ML	
08828	a	**E**	E	*EW*	CF	MG
08830	x*	**G**	CA	*SS*	CO	
08834	x	**F**	GN		BN (U)	
08836	x	**I**	GW	*GW*	OO	
08837	x*	**D**	E	*EW*	AN	
08842	x		E	*EW*	AN	
08844	x		E	*EW*	AN	Wembley
08846	x	**0**	AD	*AD*	ZC	
08847	x*		WT	*WT*	ZG	
08853	xr		GN	*GN*	BN	
08854	x*		E	*EW*	OC	
08856	x		E	*EW*	AN	
08865	x		E	*EW*	TO	PB
08866	x		E	*EW*	CD	
08867	x	**0**	E	*EW*	CD	
08868	x		MR	*JF*	ZB	Connington Tip
08869	x	**G**	AR	*AR*	NC	
08872	x	**D**	E	*EW*	AN	
08873	x	**RX**	E	*EW*	CD	Wembley
08877	x	**D**	E	*EW*	KY	
08879	x	**0**	E	*EW*	TI	
08881	x	**D**	E	*EW*	ML	
08882	x		E	*EW*	ML	AB
08883	x	**0**	E	*EW*	ML	PH
08884	x		E	*EW*	BS	
08886	x	**E**	E	*EW*	TO	
08887	x	**0**	VT	*VW*	WN	
08888	x	**E**	E	*EW*	IM	
08890	x	**D**	E	*EW*	OC	
08891	x		FL	*FL*	CD	Garston FLT
08892	x*	**GN**	RF	*GN*	BN	
08893	x	**D**	E		BS (U)	
08894	x		E		SP (U)	
08896	x	**E**	E	*EW*	BK	
08897	x	**D**	E	*EW*	CD	
08899	x	**MM**	MM	*ML*	DY	
08900	x	**D**	E	*EW*	CF	
08901	xn		E		BS (U)	
08902	x		E		AN (U)	
08903	x	**0**	I	*IC*	BH	
08904	x		E	*EW*	OC	
08905	x.	**RD**	E	*EW*	TI	SY

08906	x		E	EW	ML
08907	x	0	E	EW	AN
08908	xz		MM	ML	NL
08909	x		E	EW	BS
08910	x		E	EW	ML
08911	x	D	E	EW	CD
08912	x	BS	E	EW	ML	CL
08913	x	D	E	EW	AN	Wembley
08914	x		E		BS (U)
08915	x	F	E	EW	CD	Long Marston
08918	x	D	E		SP (U)
08919	x	RX	E	EW	BK
08920	x	F	E	EW	BS
08921	x	E	E	EW	CD
08922	x	D	E	EW	ML	CL
08924	x	D	E	EW	OC
08925	x		E		ZB (U)
08926	x		E		AN (U)
08927	x		E	EW	BS
08928	x	FR	AR	AR	NC
08931	x		E		TE (U)
08932	x		E	EW	CF
08933	x*	E	E	EW	EH
08934	x		VT	VW	WN
08938	xr	0	E		ML (U)
08939	x		E	EW	AN
08940	x		E		EH (U)
08941	x		E	EW	CF	BZ
08942	x		E	EW	CF
08943	x	0	AD	AD	ZT
08944	x	D	E	EW	OC
08946	x	RD	E	EW	TI	SY
08947	x		E	EW	OC
08948	xc	EP	LC	ES	OC
08950	x	I	MM	ML	NL
08951	x	D	E	EW	TI	SY
08952	x		E		ML (U)
08953	x	D	E	EW	CF	BZ
08954	x	FT	E	EW	CF
08955	x		E	EW	CF
08956	x		SO	SO	DY
08957	x	E	E	EW	CF
08958	x		E		SF (U)

Names:

08578	Libert Dickinson	08714	Cambridge
08649	G.H. Stratton	08743	ANGIE
08661	Europa	08790	M.A. SMITH
08682	Lionheart	08869	The Canary
08701	The Sorter	08879	Sheffield Children's Hospital
08711	EAGLE C.U.R.C.	08888	Postman's Pride

▲ Although it has been sold to Freightliner, Class 08 No. 08077 still carries the livery of its former owner, RFS. It is pictured here at Southampton Maritime Freightliner Terminal on 26th January 1997. **Brian Denton**

▼ Departmental liveried Class 09 No. 09107 passes through Newport with a Llanwern to Newport Alexandra Dock Jn transfer freight on 1st April 1997.
Bob Sweet

Direct Rail Services liveried Class 20s Nos. 20305 & 20301 'FURNESS RAILWAY 150' head south through Lowgill on 14th July 1997, with a Penrith to Cricklewood milk train.

Kevin Conkey

A pair of Class 31s, Nos. 31467, in BR blue livery, and 31229, in Civil-link livery pass Slindon, Staffordshire on 25th July 1997 with the 11.10 Sheerness–Mossend Enterprise service.

Peter Fox

▲ One of only eight Class 33s still in traffic, No. 33046 descends from Whiteball tunnel with the 10.00 Westbury–Meldon Quarry empty ballast working on 7th April 1997. The loco carries Civil-link livery. **Russell Ayre**

▼ The 13.34 Fawley–Tavistock Junction bogie tank train passes through St Denys on 22nd August 1997 with Loadhaul liveried Class 37 No. 37884 in charge.
David Brown

A pair of Transrail Class 37s Nos. 37897 & 37887 climb the bank at Stormy with eastbound steel working on 26th May 1997.
Russell Ayre

Midland Mainline liveried Class 43 No. 43085 leads a similarly liveried set forming the 11.33 Nottingham–London St Pancras at Harrowden, Wellingborough on 29th October 1997.

Michael J. Collins

Great Western Trains liveried Class 43 power cars Nos. 43132, leading, & 43185 'Great Western' provide the power for the 06.30 London Paddington–Plymouth as it passes Dawlish Warren on 20th June 1997.
C.J. Marsden

▲ The 08.11 Southampton–Ripple Lane Freightliner service enters Camden Road station behind Freightliner liveried Class 47 No. 47052 on 23rd September 1997.
Kevin Conkey

▼ Transrail class 56 No. 56070 passes through Swanley on 19th February 1996 with the 11.10 Sheerness–Willesden freight. **Rodney Lissenden**

| 08896 | Stephen Dent | 08950 | Neville Hill 1st |
| 08919 | Steep Holm | | |

Class 08/9. Fitted with cut-down cab and headlight for Cwmmawr branch.

					Depot	*Location/Sub-depot*
08993	x	**FT**	E	*EW*	CF	
08994	a	**D**	E		ZB (U)	
08995	a	**FT**	E		ZB (U)	

CLASS 09 　　　　　 BR SHUNTER 　　　 0-6-0

Built: 1959–62 by BR at Darlington or Horwich Works.
Engine: English Electric 6KT of 298 kW (400 hp) at 680 rpm.
Main Generator: English Electric 801.
Traction Motors: English Electric 506.
Max. Tractive Effort: 111 kN (25000 lbf).
Cont. Tractive Effort: 39 kN (8800 lbf) at 11.6 m.p.h.
Power At Rail: 201 kW (269 hp).
Brake Force: 19 t.
Weight: 50 t.
Max. Speed: 27 m.p.h.
Train Brakes: Air & Vacuum.

Length over Buffers: 8.92 m.
Wheel Diameter: 1372 mm.
RA: 5.

Non-standard livery:
09017 is BR blue with a grey cab and is numbered 97806.

Class 09/0 were originally numbered 3665–71, 3719–21, 4099–4114.

Class 09/0. Built as Class 09.

				Depot	*Location/Sub-depot*
09001		E	*EW*	CF	MG
09003		E	*EW*	CF	
09004		SC	*SC*	SU	
09005	**D**	E		TE (U)	
09006	**ML**	E	*EW*	HG	
09007	**ML**	E	*EW*	HG	
09008	**D**	E	*EW*	CF	Tavistock Junction
09009	**E**	E	*EW*	HG	
09010	**D**	E	*EW*	TO	PB
09011	**D**	E	*EW*	TI	SY
09012	**D**	E	*EW*	CD	
09013	**D**	E	*EW*	CF	Tavistock Junction
09014	**D**	E	*EW*	KY	
09015	**D**	E	*EW*	CF	
09016	**D**	E	*EW*	OC	
09017	**O**	E	*EW*	CF	Sudbrook
09018	**ML**	E		ZB (U)	
09019	**ML**	E	*EW*	HG	
09020		E	*EW*	OC	
09021	**RD**	E	*EW*	TI	SY

09022		E		AN (U)
09023		E		ZB (U)
09024	**ML**	E	*EW*	HG
09025		SC	*SC*	SU
09026	**D**	SC	*SC*	SU

Names:

09009	Three Bridges C.E.D.	09026	William Pearson
09012	Dick Hardy		

Class 09/1. Converted from Class 08. 110 V electrical equipment.

Depot Location/Sub-depot

09101	**D**	E	*EW*	OC	
09102	**D**	E	*EW*	OC	
09103	**D**	E	*EW*	ML	AB
09104	**D**	E	*EW*	BS	
09105	**D**	E	*EW*	CF	
09106	**D**	E	*EW*	TE	
09107	**D**	E	*EW*	CF	

Class 09/2. Converted from Class 08. 90 V electrical equipment.

Depot Location/Sub-depot

09201	**D**	E	*EW*	DR
09202	**D**	E	*EW*	ML
09203	**D**	E	*EW*	CF
09204	**D**	E	*EW*	TE
09205	**D**	E	*EW*	ML

CLASS 20 ENGLISH ELECTRIC TYPE 1 Bo-Bo

Built: 1957–68 by English Electric Company at Vulcan Foundry, Newton le Willows or Robert Stephenson & Hawthorn, Darlington.
20007–128/301–5/901–4 were originally built with disc indicators whilst 20131–215/905/6 were built with four character headcode panels.
Engine: English Electric 8SVT Mk. II of 746 kW (1000 hp) at 850 rpm.
Main Generator: English Electric 819/3C.
Traction Motors: English Electric 526/5D (20007–42/301/904) or 526/8D (others).
Max. Tractive Effort: 187 kN (42000 lbf).
Cont. Tractive Effort: 111 kN (25000 lbf) at 11 m.p.h.

Power At Rail: 574 kW (770 hp).	**Length over Buffers:** 14.25 m.
Brake Force: 35 t.	**Wheel Diameter:** 1092 mm.
Design Speed: 75 m.p.h.	**Weight:** 73.5 t.
Max. Speed: 60 m.p.h.	**RA:** 5.

Train Brakes: Air & Vacuum.
Multiple Working: Blue Star Coupling Code.

Non-standard livery:
20088, 20102, 20105, 20108, 20133, 20145, 20159 & 20194 are RFS grey with blue and yellow bodyside stripes and carry the following numbers:

20088: 2017	20105: 2016	20133: 2005	20159: 2010	
20102: 2008	20108: 2001	20145: 2019	20194: 2006	

Originally numbered in series 8007–8190, 8315–8325.

Class 20/0. Standard Design.

20007	st	D	ZB (U)
20016	st	E	BS (U)
20032	s	D	ZB (U)
20042	W	D	ZB (U)
20057	st	E	BS (U)
20059	st FR	E	MG (U)
20066		E	BS (U)
20072	st	D	ZB (U)
20075	st T	D	ZB (U)
20081	st	E	BS (U)
20087	st BS	E	BS (U)
20088	0	D	ZB (U)
20092	CS	E	BS (U)
20102	0	D	ZB (U)
20104	st FR	D	ZB (U)
20105	0	D	ZB (U)
20108	0	D	BL (U)
20117	st	D	ZB (U)
20118	FR	E	BS (U)
20121	st	D	ZB (U)
20128	st T	D	ZB (U)
20131	st T	D	ZB (U)
20132	st FR	E	BS (U)
20133	0	D	ZB (U)
20138	FR	E	BS (U)
20145	0	D	ZB (U)
20159	0	D	ZB (U)
20165	FR	E	BS (U)
20168	st	E	MG (U)
20169	st CS	E	BS (U)
20187	st T	D	ZB (U)
20190	st	D	ZB (U)
20194	0	D	ZB (U)
20209		H	ZK (U)
20215	st FR	D	ZB (U)

Class 20/3. Refurbished locos for Direct Rail Services.
All have train air brakes only, twin fuel tanks and non-standard multiple work-ing jumpers. Disc indicators or headcode panels removed.

20301	(20047)	**DR**	D	*DR*	SD	FURNESS RAILWAY 150
20302	(20084)	**DR**	D	*DR*	SD	
20303	(20127)	**DR**	D	*DR*	SD	
20304	(20120)	**DR**	D	*DR*	SD	
20305	(20095)	**DR**	D	*DR*	SD	
20306	( )					

20307	(	)
20308	(	)
20309	(	)
20310	(	)

Class 20/9. Hunslet–Barclay Ltd.
All have train air brakes only.

20901	t	**HB**	H	*HB*	ZK	NANCY
20902		**HB**	H	*HB*	ZK	LORNA
20903		**HB**	H	*HB*	ZK	ALISON
20904		**HB**	H	*HB*	ZK	JANIS
20905	t	**HB**	H	*HB*	ZK	IONA
20906		**HB**	H	*HB*	ZK	Kilmarnock 400

CLASS 31   BRUSH TYPE 2   A1A–A1A

Built: 1957–62 by Brush Traction at Loughborough.
31102/6/7/10/25/34/44/444/50/61 retain two headcode lights. Others have roof-mounted headcode boxes.
Engine: English Electric 12SVT of 1100 kW (1470 hp) at 850 rpm.
Main Generator: Brush TG160-48.
Traction Motors: Brush TM73-68.
Max. Tractive Effort: 160 kN (35900 lbf) (190 kN (42800 lbf)*).
Cont. Tractive Effort: 83 kN (18700 lbf) at 23.5 m.p.h. (99 kN (22250 lbf) at 19.7 m.p.h. *.)
Power At Rail: 872 kW (1170 hp).
Brake Force: 49 t.
Design Speed: 90 (80*) m.p.h.
Weight: 107–111 t.
RA: 5 or 6.
Length over Buffers: 17.30 m.
Driving Wheel Diameter: 1092 mm.
Centre Wheel Diameter: 1003 mm.
Train Brakes: Air & Vacuum.
ETH Index (Class 31/4): 66.
Max. Speed: 60 m.p.h. (90 m.p.h. Class 31/4).
Multiple Working: Blue Star Coupling Code.

Non-standard livery:
31116 is red, yellow, red and grey with 'Infrastructure' branding.

Originally numbered 5520–5699, 5800–5862 (not in order).

Class 31/1. Standard Design. RA5.

31102		**C**	E		CD (U)
31106	*	**C**	E		BS (U)
31107		**C**	E		BS (U)
31110		**C**	E	*EW*	BS
31113		**C**	E	*EW*	BS
31116		**0**	E		TO (U)
31119		**C**	E		CL (U)
31125		**C**	E		BS (U)
31126		**C**	E		SP (U)
31128		**F0**	E		BS (U)
31130		**F**	E		BS (U)
31132		**F0**	E		BS (U)

31134		C	E		SP (U)
31135		C	E		TO (U)
31142		C	E	*EW*	BS
31144		C	E		CL (U)
31145		C	E		SP (U)
31146	r	C	E	*EW*	BS	Brush Veteran
31147	r	C	E		BS (U)
31149		**FR**	E		TO (U)
31154		C	E	*EW*	BS
31155		F	E		BS (U)
31158		C	E		BS (U)
31160		F	E		SP (U)
31163		C	E	*EW*	BS
31164		**FO**	E		BS (U)
31165		**G**	E		TO (U)
31166	r	C	E	*EW*	BS
31171		**FO**	E		BS (U)
31174		C	E		BS (U)
31178		C	E		BS (U)
31181		C	E		TO (U)
31185		C	E		BS (U)
31186		C	E		TO (U)
31187		C	E		TO (U)
31188		C	E	*EW*	BS
31190		C	E		CL (U)
31191		C	E		TO (U)
31199		F	E		TO (U)
31200		F	E		CD (U)
31201		F	E	*EW*	BS
31203		C	E	*EW*	BS
31205		**FR**	E		TO (U)
31206		C	E		BS (U)
31207		C	E	*EW*	BS
31219		C	E		TO (U)
31224		C	E		CL (U)
31229		C	E	*EW*	BS
31230	*	**FO**	E		TO (U)
31232		C	E		BS (U)
31233		C	E	*EW*	BS	Severn Valley Railway
31235		C	E		CL (U)
31237		C	E		BS (U)
31238		C	E		SP (U)
31242		C	E		CL (U)
31247		**FR**	E		TO (U)
31248		**FO**	E		BS (U)
31250		C	E		TO (U)
31252		**FO**	E		PB (U)
31255		C	E	*EW*	BS
31263		C	E		BS (U)
31268		C	E		TO (U)
31270		F	E		CL (U)

31271	F	E		TO (U)
31273	C	E	*EW*	BS
31275	F	E		CN (U)
31276	F	E		TO (U)
31285	C	E		CL (U)
31294	F	E		TO (U)
31301	FR	E		BS (U)
31302	F	E		SP (U)
31304	F	E		SP (U)
31306	C	E	*EW*	BS
31308	C	E	*EW*	BS
31312	F	E		SP (U)
31317	FO	E		BS (U)
31319	F	E		CD (U)
31327	FR	E		CL (U)

Class 31/4. Equipped with Train Heating. RA6.
Class 31/5. Train Heating Equipment isolated. RA6.

31405	M	E		TO (U)	
31407	ML	E	*EW*	BS	
31408		E		SP (U)	
31410	RR	E		CN (U)	
31411	D	E		BS (U)	
31512	C	E	*EW*	BS	
31514	C	E		BS (U)	
31415		E		BS (U)	
31516	C	E		BS (U)	
31417	D	E		BS (U)	
31519	C	E		SP (U)	
31420	M	E	*EW*	BS	
31421	RR	E		CD (U)	
31422	M	E		BS (U)	
31423	M	E		BS (U)	
31524	C	E		BS (U)	
31526	C	E		BS (U)	
31427		E		CL (U)	
31530	C	E	*EW*	BS	
31531	C	E		TO (U)	
31432		E		SP (U)	
31533	C	E		BS (U)	
31434		E	*EW*	BS	
31435	C	E		BS (U)	
31537	C	E		BS (U)	
31538		E		CL (U)	
31439	RR	E		BS (U)	North Yorkshire Moors Railway
31541	C	E		HG (U)	
31444	C	E		SP (U)	
31545		E		BS (U)	
31546	C	E		BS (U)	
31548	C	E		BS (U)	
31549	C	E		TO (U)	

31450		E	*EW*	BS
31551	C	E		TO (U)
31552	C	E		TO (U)
31554	C	E	*EW*	BS
31455	RR	E		SP (U)
31556	C	E		CL (U)
31558	C	E		TO (U)
31459		E		TO (U)
31461	D	E		TO (U)
31462	D	E		BS (U)
31563	C	E		TO (U)
31465	RR	E	*EW*	BS
31466	C	E	*EW*	BS
31467		E	*EW*	BS
31468	C	E		TO (U)

CLASS 33   BRCW TYPE 3   Bo–Bo

Built: 1960–62 by the Birmingham Railway Carriage & Wagon Company, Smethwick.
Engine: Sulzer 8LDA28 of 1160 kW (1550 hp) at 750 rpm.
Main Generator: Crompton Parkinson CG391B1.
Traction Motors: Crompton Parkinson C171C2.
Max. Tractive Effort: 200 kN (45000 lbf).
Cont. Tractive Effort: 116 kN (26000 lbf) at 17.5 m.p.h.
Power At Rail: 906 kW (1215 hp).   **Length over Buffers:** 15.47 m.
Brake Force: 35 t.   **Wheel Diameter:** 1092 mm.
Design Speed: 85 m.p.h.   **Weight:** 77.5 t (78.5 t Class 33/1).
Max. Speed: 60 m.p.h.   **RA:** 6.
Train Heating: Electric (y isolated).   **ETH Index:** 48.
Train Brakes: Air & Vacuum (Class 33/1 also has electro-pneumatic).
Multiple Working: Blue Star Coupling Code.

Non-standard livery:
33021 is Post Office red.

33021 is owned by Alan & Tracey Lear and managed by Fragonset Railways.
33116 carries its original number D 6535.
33208 carries its original number D 6593.

Originally numbered in series 6500–97 but not in order.

Class 33/0. Standard Design.

33019		C	E	*EW*	EH
33021		0	FG		TM (U) Eastleigh
33025		C	E	*EW*	EH
33026		C	E	*EW*	EH
33030		C	E	*EW*	EH
33046	y	C	E	*EW*	EH
33051			E	*EW*	EH Shakespeare Cliff

Class 33/1. Fitted with Buckeye Couplings & SR Multiple Working Equipment for use with SR EMUs, TC stock & Class 73.
Also fitted with flashing light adaptor for use on Weymouth Quay line.

33116		E	*EW*	EH	Hertfordshire Rail Tours

Class 33/2. Built to Former Loading Gauge of Tonbridge–Battle Line.
All equipped with slow speed control.

33202	y	**C**	E	*EW*	EH
33208		**G**	MH	*CA*	RL

CLASS 37   ENGLISH ELECTRIC TYPE 3   Co–Co

Built: 1960–5 by English Electric Company at Vulcan Foundry, Newton le Willows or Robert Stephenson & Hawthorn, Darlington.
37003–115/340/1/3/350/1/9 with the exception of 37019*/047/065*/072*/073/074/075*/100* (* one end only) retain box-type route indicators, the remainder having central headcode panels/marker lamps.
Engine: English Electric 12CSVT of 1300 kW (1750 hp) at 850 rpm.
Main Generator: English Electric 822/10G.
Traction Motors: English Electric 538/A.
Max. Tractive Effort: 245 kN (55500 lbf).
Cont. Tractive Effort: 156 kN (35000 lbf) at 13.6 m.p.h.

Power At Rail: 932 kW (1250 hp).	**Length over Buffers:** 18.75 m.
Brake Force: 50 t.	**Wheel Diameter:** 1092 mm.
Design Speed: 90 m.p.h.	**Weight:** 103–108 t.
Max. Speed: 80 m.p.h.	**RA:** 5 or 7.

Train Heating: Electric (Class 37/4 only). **ETH Index:** 30.
Train Brakes: Air & Vacuum.
Multiple Working: Blue Star Coupling Code.

Non-standard livery:
37116 is BR blue with Transrail markings.

a Vacuum brake isolated.

Originally numbered 6600–8, 6700–6999 (not in order). 37274 is the second loco to carry that number. It was renumbered to avoid confusion with Class 37/3 locos.

Class 37/0. Unrefurbished Locos. Technical details as above. RA5.

37003	+	**C**	E		IM (U)
37010		**C**	E	*EW*	TO
37012		**C**	E	*EW*	TO
37013	+	**ML**	E	*EW*	TO
37019	+	**F**	E		HM (U)
37023		**ML**	E	*EW*	TO Stratford TMD Quality Approved
37025		**BR**	E	*EW*	EH Inverness TMD Quality Assured
37026	+	**F**	E		SP (U)
37035		**C**	E		TO (U)
37037		**FM**	E	*EW*	EH
37038		**C**	E	*EW*	TO

37040		**E**	E	*EW*	EH
37042	+	**E**	E	*EW*	TO
37043		**CT**	E	*EW*	ML
37045	+	**F**	E		TO (U)
37046		**C**	E	*EW*	IM
37047	+	**ML**	E	*EW*	EH
37048		**FM**	E		TO (U)
37051		**E**	E	*EW*	TO	Merehead
37054		**C**	E	*EW*	EH
37055		**ML**	E	*EW*	TO	RAIL Celebrity
37057	+	**E**	E	*EW*	TO	Viking
37058	+	**C**	E	*EW*	IM
37059	+	**F**	E		IM (U)
37063	+	**F**	E		TE (U)
37065	+	**ML**	E	*EW*	EH
37068	+	**F**	E		IM (U)
37069	+	**C**	E	*EW*	ML
37071	+	**C**	E	*EW*	IM
37072	+	**D**	E		IM (U)
37073	+	**FT**	E	*EW*	TO	Fort William/An Gearasdan
37074	+	**ML**	E	*EW*	EH
37075	+	**F**	E		IM (U)
37077		**ML**	E	*EW*	EH
37078	+	**F**	E		ML (U)
37079	+	**F**	E	*EW*	TO
37083	+	**C**	E		IM (U)
37087		**C**	E		CD (U)
37088		**CT**	E		ML (U)	Clydesdale
37092		**C**	E		TO (U)
37095	+	**C**	E	*EW*	TO
37097		**C**	E	*EW*	EH
37098	+	**C**	E	*EW*	IM
37100	+	**FT**	E	*EW*	ML
37101	+	**F**	E		IM (U)
37104		**C**	E		IM (U)
37106	+	**C**	E	*EW*	EH
37107	+	**F**	E		SP (U)
37108	+	**F**	E		BS (U)
37109		**E**	E	*EW*	EH
37110	+	**F**	E		IM (U)
37111		**FT**	E		TO (U)
37114	+	**E**	E	*EW*	TO	City of Worcester
37116	+	**O**	E	*EW*	EH	Sister Dora
37131	+	**F**	E	*EW*	IM
37133		**C**	E	*EW*	TO
37137		**FM**	E		TO (U)
37139	+	**F**	E		TE (U)
37140		**C**	E	*EW*	EH
37141		**C**	E		IM (U)
37142		**C**	E		CD (U)
37144	r	**F**	E		IM (U)

37146		C	E	*EW*	TO
37152		I	E	*EW*	ML
37153		CT	E	*EW*	ML
37154	+	FT	E	*EW*	TO
37156	r	FT	E	*EW*	IM
37158		C	E	*EW*	TO
37162	+	D	E	*EW*	TO
37165	+	C	E	*EW*	ML
37170	r	C	E	*EW*	ML
37174		E	E	*EW*	EH
37175		C	E	*EW*	ML
37178	+	F	E	*EW*	CF
37184		C	E		BS (U)
37185	+	C	E	*EW*	TO	Lea & Perrins
37188		C	E		TO (U)
37191		C	E	*EW*	TO
37194	+	FM	E	*EW*	EH	British International Freight Association
37196		C	E	*EW*	TO
37197	+	CT	E	*EW*	CF
37198	+	ML	E	*EW*	EH
37201		CT	E		BS (U)
37203		ML	E	*EW*	EH
37207		C	E		BS (U)
37209		BR	E		DR (U)
37211		C	E	*EW*	EH
37212	+	FT	E	*EW*	IM
37213	+	F	E		TO (U)
37214	+	FT	E		BS (U)
37216	r+	ML	E	*EW*	TO	Great Eastern
37217	+		E		IM (U)
37218	+	F	E		IM (U)
37219	r	ML	E	*EW*	EH
37220	+	E	E	*EW*	TO
37221		FT	E	*EW*	ML
37222	+	FM	E		CF (U)
37223	+	F	E		IM (U)
37225	+	F	E	*EW*	IM
37227	+	FM	E		SL (U)
37229	+	F	E	*EW*	CF
37230	+	CT	E	*EW*	CF
37232	r	CT	E		ML (U)	The Institution of Railway Signal Engineers
37235	+	F	E		DR (U)
37238	+	F	E	*EW*	TO
37240	+	C	E		BS (U)
37241		F	E		TO (U)
37242	+	ML	E	*EW*	EH
37244	+	F	E	*EW*	IM
37245		C	E	*EW*	EH
37248	+	ML	E	*EW*	TO	Midland Railway Centre

37250	+	FT	E	EW	ML	
37251	+	I	E		ML (U)	
37254	+	C	E	EW	CF	
37255	+	C	E	EW	TO	
37258	+	C	E	EW	TO	
37261	+	F	E	EW	ML	Caithness
37262	+	D	E	EW	EH	
37263		C	E	EW	CF	
37264		C	E	EW	TO	
37274	+	ML	E	EW	EH	
37275	+		E	EW	CF	Oor Wullie
37278	+	F	E		TO (U)	
37293	+	ML	E	EW	EH	
37294	+	C	E	EW	ML	
37298	+	F	E		IM (U)	

Class 37/3. Unrefurbished locos fitted with regeared (CP7) bogies. Details as Class 37/0 except:

Max. Tractive Effort: 250 kN (56180 lbf).
Cont. Tractive Effort: 184 kN (41250 lbf) at 11.4 m.p.h.

37330	(37128)	+	BR	E		TO (U)	
37331	(37202)		F	E		DR (U)	
37332	(37239)	+	F	E	EW	TO	
37334	(37272)	+	F	E		IM (U)	
37335	(37285)	+	F	E		IM (U)	
37340	(37009)	+	F	E		IM (U)	
37341	(37015)	+	F	E		TE (U)	
37343	(37049)		C	E		TO (U)	
37344	(37053)	+	F	E		IM (U)	
37350	(37119)	+	F	E	EW	IM	
37351	(37002)	+	CT	E	EW	ML	
37358	(37091)		F	E		IM (U)	
37359	(37118)		F	E		TE (U)	
37370	(37127)		E	E	EW	EH	
37371	(37147)	+	ML	E	EW	EH	
37372	(37159)		ML	E	EW	EH	
37375	(37193)	+	ML	E	EW	EH	
37376	(37199)	+	F	E	EW	IM	
37377	(37200)	+	C	E	EW	EH	
37379	(37226)		ML	E	EW	TO	Ipswich WRD Quality Assured
37380	(37259)		FM	E	EW	EH	
37381	(37284)	+	F	E		FH (U)	
37382	(37145)		F	E		IM (U)	
37383	(37167)	+	ML	E	EW	EH	
37384	( )						

Class 37/4. Refurbished locos fitted with train heating. Main generator replaced by alternator. Regeared (CP7) bogies. Details as class 37/0 except:

Main Alternator: Brush BA1005A.
Max. Tractive Effort: 256 kN (57440 lbf).
Cont. Tractive Effort: 184 kN (41250 lbf) at 11.4 m.p.h.
Power At Rail: 935 kW (1254 hp).
All have twin fuel tanks.

37401	r	**FT**	E	*EW*	ML	Mary Queen of Scots
37402	r	**F**	E	*EW*	IM	Bont Y Bermo
37403	r	**G**	E	*EW*	TO	Ben Cruachan
37404	r	**FT**	E	*EW*	ML	Loch Long
37405	r	**E**	E	*EW*	ML	
37406	r	**FT**	E	*EW*	ML	The Saltire Society
37407	r	**FT**	E	*EW*	IM	Blackpool Tower
37408		**BR**	E	*EW*	IM	Loch Rannoch
37409	r	**FT**	E	*EW*	ML	Loch Awe
37410	r	**FT**	E	*EW*	ML	Aluminium 100
37411		**E**	E	*EW*	CF	Ty Hafan
37412		**FT**	E	*EW*	CF	Driver John Elliot
37413	r	**E**	E	*EW*	ML	The Scottish Railway Preservation Society
37414	r	**RR**	E	*EW*	CD	Cathays C&W Works 1846–1993
37415	r	**E**	E	*EW*	CD	
37416	r	**E**	E	*EW*	CF	
37417	r	**F**	E	*EW*	IM	Highland Region
37418	r	**E**	E	*EW*	CD	East Lancashire Railway
37419	r	**E**	E	*EW*	CD	
37420	r	**RR**	E	*EW*	CD	The Scottish Hosteller
37421	r	**E**	E	*EW*	CD	
37422	r	**RR**	E	*EW*	CD	Robert F. Fairlie Locomotive Engineer 1831–1885
37423	r	**FT**	E	*EW*	TO	Sir Murray Morrison 1873–1948 Pioneer of British Aluminium Industry
37424	r	**FT**	E	*EW*	ML	
37425	r	**RR**	E	*EW*	ML	Sir Robert McAlpine/ Concrete Bob (opposite sides)
37426	r	**E**	E	*EW*	CD	
37427	r	**E**	E	*EW*	CF	
37428	r	**F**	E	*EW*	ML	David Lloyd George
37429	r	**RR**	E	*EW*	IM	Eisteddfod Genedlaethol
37430	r	**FT**	E	*EW*	ML	Cwmbrân
37431	r	**M**	E	*EW*	TO	

Class 37/5. Refurbished locos. Main generator replaced by alternator. Regeared (CP7) bogies. Details as class 37/4 except:

Max. Tractive Effort: 248 kN (55590 lbf).
All have twin fuel tanks.

37503		**E**	E	*EW*	TO	
37505		**FT**	E	*EW*	IM	British Steel Workington

37509		F	E	*EW*	IM
37510		I	E	*EW*	ML
37513		**LH**	E	*EW*	IM
37515	s	**E**	E	*EW*	IM
37516	s	**LH**	E	*EW*	IM
37517	ars	**E**	E	*EW*	IM
37518		**E**	E	*EW*	IM
37519		F	E	*EW*	TO
37520		**E**	E	*EW*	ML
37521		**E**	E	*EW*	CF	English China Clays

Class 37/6. Refurbished locos modified for use with Nightstar stock. Main generator replaced by alternator. Class 50 bogies. Details as class 37/0 except:

Main Alternator: Brush BA1005A.
All have twin fuel tanks, train air brakes only, UIC brake and coaching stock jumpers, RCH jumpers, ETH through wires.

37601	(37501)	**EP**	LC	*ES*	OC
37602	(37502)	**EP**	LC	*ES*	OC
37603	(37504)	**EP**	LC	*ES*	OC
37604	(37506)	**EP**	LC	*ES*	OC
37605	(37507)	**EP**	LC	*ES*	OC
37606	(37508)	**EP**	LC	*ES*	OC
37607	(37511)	**EP**	D	*DR*	SD
37608	(37512)	**DR**	D	*DR*	SD
37609	(37514)	**DR**	D	*DR*	SD
37610	(37687)	**EP**	D	*DR*	SD
37611	(37690)	**DR**	D	*DR*	SD
37612	(37691)	**EP**	D	*DR*	SD

Class 37/5 continued.

37667	s	**E**	E	*EW*	ML	Meldon Quarry Centenary
37668	s	**E**	E	*EW*	CF	
37669		**E**	E	*EW*	CF	
37670		**E**	E	*EW*	CF	
37671		**FT**	E	*EW*	CF	Tre Pol and Pen
37672	s	**F**	E	*EW*	CF	
37673		**FT**	E	*EW*	CF	
37674		**FT**	E	*EW*	CF	Saint Blaise Church 1445–1995
37675	s	**FT**	E	*EW*	IM	
37676		**F**	E	*EW*	IM	
37677		**F**	E	*EW*	TO	
37678		**F**	E	*EW*	IM	
37679		**F**	E	*EW*	IM	
37680		**F**	E	*EW*	TO	
37682	r	**E**	E	*EW*	IM	Hartlepool Pipe Mill
37683		**FT**	E	*EW*	IM	
37684		**E**	E	*EW*	ML	Peak National Park
37685		**I**	E	*EW*	IM	
37686		**F**	E	*EW*	IM	

37688		E	E	EW	IM
37689	s	F	E	EW	IM
37692	s	F	E	EW	ML	The Lass O' Ballochmyle
37693	s	FT	E	EW	ML
37694		E	E	EW	IM
37695	s	E	E	EW	IM
37696	s	FT	E	EW	CF
37697	s	E	E	EW	IM
37698	s	LH	E	EW	IM

Class 37/7. Refurbished locos. Main generator replaced by alternator. Regeared (CP7) bogies. Ballast weights added. Details as class 37/4 except:

Main Alternator: GEC G564AZ (37796–803) Brush BA1005A (others).
Max. Tractive Effort: 276 kN (62000 lbf).
Weight: 120 t.   **RA:** 7.
All have twin fuel tanks.

37701	s	FT	E	EW	CF
37702	s	FT	E	EW	ML	Taff Merthyr
37703	s	E	E	EW	EH
37704	s	E	E	EW	CF
37705		FM	E	EW	EH
37706		E	E	EW	TO
37707		E	E	EW	IM
37708		F	E	EW	IM
37709		FM	E	EW	EH
37710		LH	E	EW	IM
37711		E	E	EW	EH
37712		E	E	EW	ML
37713		LH	E	EW	IM
37714		E	E	EW	ML
37715		FM	E	EW	TO	British Petroleum
37716		E	E	EW	IM
37717		E	E	EW	IM	St Margaret's Church of England Primary School City of Durham Railsafe Trophy Winners 1997
37718		E	E	EW	IM
37719		F	E	EW	IM
37796	s	E	E	EW	ML
37797	s	E	E	EW	ML
37798	s	ML	E	EW	TO
37799	s	FT	E	EW	ML	Sir Dyfed/County of Dyfed
37800	s	E	E	EW	EH
37801	s	E	E	EW	ML
37802	s	FT	E	EW	ML
37803	s	ML	E	EW	EH
37883		E	E	EW	IM
37884		LH	E	EW	IM	Gartcosh
37885		E	E	EW	IM
37886		E	E	EW	IM
37887	s	FT	E	EW	CF

37888		F	E	*EW*	CF	
37889		FT	E	*EW*	CF	
37890	a	FM	E	*EW*	EH	The Railway Observer
37891		FM	E	*EW*	EH	
37892		FM	E	*EW*	EH	Ripple Lane
37893		E	E	*EW*	ML	
37894	s	E	E	*EW*	CF	
37895	s	E	E	*EW*	CF	
37896	s	FT	E	*EW*	CF	
37897	s	FT	E	*EW*	CF	
37898	s	FT	E	*EW*	CF	Cwmbargoed DP
37899	s	E	E	*EW*	TO	

Class 37/9. Refurbished Locos. Fitted with manufacturers prototype power units and ballast weights. Main generator replaced by alternator. Details as Class 37/0 except:

Engine: Mirrlees MB275T of 1340 kW (1800 hp) at 1000 rpm (37901–4), Ruston RK270T of 1340 kW (1800 hp) at 900 rpm (37905/6).
Main Alternator: Brush BA1005A (GEC G564, 37905/6).
Max. Tractive Effort: 279 kN (62680 lbf).
Cont. Tractive Effort: 184 kN (41250 lbf) at 11.4 m.p.h.
Weight: 120 t.   **RA:** 7.
All have twin fuel tanks.

37901		FT	E	*EW*	CF	Mirrlees Pioneer
37902		F	E	*EW*	CF	
37903		F	E	*EW*	CF	
37904		F	E		CF (U)	
37905	s	F	E	*EW*	CF	
37906	s	FT	E		CF (U)	

CLASS 43   HST POWER CAR   Bo–Bo

Built: 1976–82 by BREL Crewe Works. Formerly numbered as coaching stock but now classified as locomotives. Fitted with luggage compartment.
Engine: Paxman Valenta 12RP200L (Paxman VP185*) of 1680 kW (2250 hp) at 1500 rpm.
Main Alternator: Brush BA1001B.
Traction Motors: Brush TMH68–46 or GEC G417AZ (43124–151/180). Frame mounted.
Max. Tractive Effort: 80 kN (17980 lbf).
Cont. Tractive Effort: 46 kN (10340 lbf) at 64.5 m.p.h.
Power At Rail: 1320 kW (1770 hp).   **ETH:** Non standard 3-phase system.
Brake Force: 35 t.   **Length over Buffers:** 17.79 m.
Weight: 70 t.   **Wheel Diameter:** 1020 mm.
Max. Speed: 125 m.p.h.   **RA:** 5.
Train Brakes: Air.
Multiple Working: With one other similar vehicle.
Communication Equipment: All equipped with driver–guard telephone.

43002	I	A	GW	PM	
43003	GW	A	GW	PM	
43004	GW	A	GW	PM	Borough of Swindon
43005	GW	A	GW	PM	
43006	I	A	GW	LA	
43007	I	A	GW	LA	
43008	GW	A	GW	LA	
43009	GW	A	GW	PM	
43010	GW	A	GW	PM	
43011	GW	A	GW	PM	Reader 125
43012	GW	A	GW	PM	
43013	I	P	VX	EC	CROSSCOUNTRY VOYAGER
43014	I	P	VX	EC	
43015	GW	A	GW	PM	
43016	GW	A	GW	PM	
43017	GW	A	GW	LA	
43018	GW	A	GW	LA	The Red Cross
43019	GW	A	GW	LA	City of Swansea/Dinas Abertawe
43020	GW	A	GW	LA	John Grooms
43021	I	A	GW	LA	
43022	GW	A	GW	LA	
43023	GW	A	GW	LA	County of Cornwall
43024	GW	A	GW	LA	
43025	I	A	GW	LA	Exeter
43026	GW	A	GW	LA	City of Westminster
43027	I	A	GW	LA	Glorious Devon
43028	I	A	VW	LO	
43029	I	A	VW	LO	
43030	GW	A	GW	PM	
43031	GW	A	GW	PM	
43032	GW	A	GW	PM	The Royal Regiment of Wales
43033	I	A	GW	PM	
43034	GW	A	GW	PM	The Black Horse
43035	I	A	GW	PM	
43036	GW	A	GW	PM	
43037	I	A	GW	PM	
43038	GN	A	GN	NL	
43039	GN	A	GN	NL	
43040	I	A	GW	PM	
43041	I	A	VW	LO	City of Discovery
43042	I	A	VW	LO	
43043	MM	P	ML	NL	LEICESTERSHIRE COUNTY CRICKET CLUB
43044	MM	P	ML	NL	Borough of Kettering
43045	MM	P	ML	NL	
43046	MM	P	ML	NL	Royal Philharmonic
43047 *	MM	P	ML	NL	
43048	I	P	ML	NL	
43049	MM	P	ML	NL	Neville Hill
43050	I	P	ML	NL	
43051	I	P	ML	NL	The Duke and Duchess of York

43052	I	P	*ML*	NL	City of Peterborough
43053	I	P	*ML*	NL	Leeds United
43054	I	P	*ML*	NL	
43055	MM	P	*ML*	NL	Sheffield Star
43056	MM	P	*ML*	NL	
43057	I	P	*ML*	NL	Bounds Green
43058	MM	P	*ML*	NL	
43059	* MM	P	*ML*	NL	MIDLAND PRIDE
43060	MM	P	*ML*	NL	County of Leicestershire
43061	MM	P	*ML*	NL	
43062	I	P	*VX*	EC	
43063	V	P	*VX*	EC	Maiden Voyager
43064	I	P	*ML*	NL	City of York
43065	I	P	*VX*	EC	City of Edinburgh
43066	MM	P	*ML*	NL	Nottingham Playhouse
43067	I	P	*VX*	EC	
43068	V	P	*VX*	EC	The Red Nose
43069	I	P	*VX*	EC	
43070	I	P	*VX*	EC	
43071	I	P	*VX*	EC	Forward Birmingham
43072	MM	P	*ML*	NL	Derby Etches Park
43073	I	P	*ML*	NL	
43074	* MM	P	*ML*	NL	BBC EAST MIDLANDS TODAY
43075	* I	P	*ML*	NL	
43076	MM	P	*ML*	NL	THE MASTER CUTLER 1947-1997
43077	MM	P	*ML*	NL	
43078	I	P	*VX*	EC	Golowan Festival Penzance
43079	I	P	*VX*	EC	
43080	I	P	*VX*	EC	
43081	MM	P	*ML*	NL	
43082	MM	P	*ML*	NL	DERBYSHIRE FIRST
43083	MM	P	*ML*	NL	
43084	V	P	*VX*	EC	County of Derbyshire
43085	MM	P	*ML*	NL	
43086	I	P	*VX*	EC	
43087	I	P	*VX*	LA	
43088	I	P	*VX*	LA	XIII Commonwealth Games Scotland 1986
43089	I	P	*VX*	LA	
43090	V	P	*VX*	LA	
43091	I	P	*VX*	LA	Edinburgh Military Tattoo
43092	V	P	*VX*	EC	Institution of Mechanical Engineers 150th Anniversary 1847-1997
43093	V	P	*VX*	EC	Lady in Red
43094	I	P	*VX*	EC	
43095	GN	A	*GN*	NL	
43096	GN	A	*GN*	NL	The Great Racer
43097	I	P	*VX*	EC	
43098	V	P	*VX*	EC	
43099	I	P	*VX*	EC	
43100	I	P	*VX*	EC	Craigentinny

43101	I	P	VX	LA	Edinburgh International Festival
43102	I	P	VX	LA	
43103	I	P	VX	LA	John Wesley
43104	I	A		LA (U)	County of Cleveland
43105	GN	A	GN	NL	
43106	GN	A	GN	NL	
43107	GN	A	GN	NL	
43108	GN	A	GN	NL	
43109	GN	A	GN	NL	
43110	GN	A	GN	EC	
43111	GN	A	GN	EC	
43112	GN	A	GN	EC	
43113	GN	A	GN	EC	
43114	GN	A	GN	EC	
43115	GN	A	GN	EC	
43116	GN	A	GN	EC	
43117	GN	A	GN	EC	
43118	GN	A	GN	EC	
43119	GN	A	GN	EC	
43120	GN	A	GN	EC	
43121	I	P	VX	LA	West Yorkshire Metropolitan County
43122	I	P	VX	LA	South Yorkshire Metropolitan County
43123	I	P	VX	EC	
43124	GW	A	GW	PM	
43125	GW	A	GW	PM	Merchant Venturer
43126	I	A	GW	PM	City of Bristol
43127	I	A	GW	PM	
43128	GW	A	GW	PM	
43129	GW	A	GW	PM	
43130	I	A	GW	PM	Sulis Minerva
43131	GW	A	GW	PM	Sir Felix Pole
43132	GW	A	GW	PM	
43133	I	A	GW	PM	
43134	I	A	GW	PM	County of Somerset
43135	GW	A	GW	PM	
43136	GW	A	GW	PM	
43137	GW	A	GW	PM	Newton Abbot 150
43138	GW	A	GW	PM	
43139	GW	A	GW	PM	
43140	GW	A	GW	PM	
43141	GW	A	GW	PM	
43142	GW	A	GW	PM	
43143	I	A	GW	PM	
43144	I	A	GW	PM	
43145	I	A	GW	PM	
43146	I	A	GW	PM	
43147	I	A	GW	PM	
43148	I	A	GW	PM	
43149	GW	A	GW	PM	BBC Wales Today
43150	I	A	GW	PM	Bristol Evening Post
43151	I	A	GW	PM	

43152	**I**	A	*GW*	PM	
43153	**V**	P	*VX*	LA	THE ENGLISH RIVIERA TORQUAY PAIGNTON BRIXHAM
43154	**V**	P	*VX*	LA	INTERCITY
43155	**V**	P	*VX*	LA	The Red Arrows
43156	**I**	P	*VX*	LA	
43157	**I**	P	*VX*	LA	Yorkshire Evening Post
43158	**I**	P	*VX*	LA	Dartmoor The Pony Express
43159	**I**	P	*VX*	LA	
43160	**V**	P	*VX*	LA	
43161	**I**	P	*VX*	LA	Reading Evening Post
43162	**I**	P	*VX*	LA	Borough of Stevenage
43163	**I**	A	*GW*	LA	
43164	**I**	A	VW	LO	
43165	**I**	A	VW	LO	
43166	**I**	A	VW	LO	
43167	*	**GN**	A	*GN*	NL	
43168	*	**GW**	A	*GW*	LA	
43169	*	**GW**	A	*GW*	LA	The National Trust
43170	*	**GW**	A	*GW*	LA	Edward Paxman
43171	**I**	A	*GW*	LA	
43172	**I**	A	*GW*	LA	
43173	*	**GW**	A		ZC (U)	
43174	**GW**	A	*GW*	LA	Bristol - Bordeaux
43175	*	**I**	A	*GW*	LA	
43176	**I**	A	*GW*	LA	
43177	*	**GW**	A	*GW*	LA	University of Exeter
43178	**GW**	A	*GW*	LA	
43179	*	**GW**	A	*GW*	LA	Pride of Laira
43180	**I**	P	*GW*	LA	
43181	**I**	A	*GW*	LA	Devonport Royal Dockyard 1693-1993
43182	**GW**	A	*GW*	LA	
43183	**GW**	A	*GW*	LA	
43184	**I**	A	*GW*	LA	
43185	**GW**	A	*GW*	LA	Great Western
43186	**GW**	A	*GW*	LA	Sir Francis Drake
43187	**GW**	A	*GW*	LA	
43188	**GW**	A	*GW*	LA	City of Plymouth
43189	**GW**	A	*GW*	LA	RAILWAY HERITAGE TRUST
43190	**GW**	A	*GW*	LA	
43191	*	**GW**	A	*GW*	LA	Seahawk
43192	**GW**	A	*GW*	LA	City of Truro
43193	**I**	P	*VX*	LA	Plymouth SPIRIT OF DISCOVERY
43194	**I**	P	*VX*	LA	
43195	**I**	P	*VX*	LA	British Red Cross 125th Birthday 1995
43196	**I**	P	*VX*	LA	The Newspaper Society Founded 1836
43197	**I**	P	*VX*	LA	Railway Magazine 1897 Centenary 1997
43198	**I**	P	*VX*	LA	

CLASS 46   BR TYPE 4   1Co-Co1

Built: 1962 by BR Derby Locomotive Works.
Engine: Sulzer 12LDA28B of 1860 kW (2500 hp) at 750 rpm.
Main Generator: Brush TG160-60.
Traction Motors: Brush TM73-68 Mk3 (axle hung).
Max. Tractive Effort: 245 kN (55000 lbf).
Cont. Tractive Effort: 141 kN (31600 lbf) at 22.3 m.p.h.
Power At Rail: 1460 kW (1960 hp).   **Length over Buffers:** 20.70 m.
Brake Force: 63 t.   **Wheel Diameter:** 914/1143 mm.
Design Speed: 90 m.p.h.   **Weight:** 141 t.
Max. Speed: 75 m.p.h.   **RA:** 7.
Train Brakes: Air & Vacuum.   **Multiple Working:** Not equipped.

Carries original number D 172.

46035	**G**	LN	*SS*	CQ	Ixion

CLASS 47   BRUSH TYPE 4   Co-Co

Built: 1963–67 by Brush Traction, Loughborough or BR Crewe Works.
Engine: Sulzer 12LDA28C of 1920 kW (2580 hp) at 750 rpm.
Main Generator: Brush TG160-60 Mk2, TG160-60 Mk4 or TM172-50 Mk1.
Traction Motors: Brush TM64-68 Mk1 or Mk1A (axle hung).
Max. Tractive Effort: 267 kN (60000 lbf).
Cont. Tractive Effort: 133 kN (30000 lbf) at 26 m.p.h.
Power At Rail: 1550 kW (2080 hp).   **Length over Buffers:** 19.38 m.
Brake Force: 61 t.   **Wheel Diameter:** 1143 mm.
Design Speed: 95 m.p.h.   **Weight:** 120.5–125 t.
Max. Speed: various.   **RA:** 6 or 7.
Train Brakes: Air & Vacuum.
Multiple Working: Green Circle (m) or Blue Star (*) Coupling Code. Otherwise not equipped.
ETH Index (47/4, 47/6 and 47/7): 66 (75 Class 47/6).

Non-standard liveries:
47114 is two-tone green with Freightliner lettering and markings.
47145 is dark blue with Railfreight Distribution markings.
47627 is maroon.
47798 & 47799 are royal train purple with a maroon and gold bodyside stripes.
47803 is grey, red and yellow.
47846 is white.

47701 is owned by Alan & Tracey Lear and managed by Fragonset Railways.
a Vacuum brake isolated.

Formerly numbered 1100–11, 1500–1999 not in order.

Class 47/0. Built with train heating boiler. RA6. Max. Speed 75 m.p.h.

47004	**G**	E	*EW*	IM	Old Oak Common Traction & Rolling Stock Depot
47016	**FO**	E	*EW*	IM	ATLAS

Number						Name
47033 am+	**RD**	E	*EW*	TI	The Royal Logistics Corps
47049 am+	**RD**	E	*EW*	TI	GEFCO
47051 am+	**RD**	E	*EW*	TI	
47052	**FL**	P	*FL*	CD	
47053 am+	**RD**	E	*EW*	TI	Dollands Moor International
47060 a	**F**	P	*FL*	CD	
47079	**FL**	FL	*FL*	CD	
47085 am+	**RD**	E	*EW*	TI	REPTA 1893–1993
47095 am+	**RD**	E	*EW*	TI	
47114 am+	**O**	FL	*FL*	CD	Freightlinerbulk
47125 am+	**RD**	E		TI (U)	
47142	**FR**	P		BL (U)	
47144 am+	**F**	E		TI (U)	
47145 am+	**O**	E	*EW*	TI	
47146 am	**RD**	E	*EW*	TI	Loughborough Grammar School
47147	**F**	P		BL (U)	
47150 am+	**RD**	FL	*FL*	CD	
47152 am+	**FL**	FL	*FL*	CD	
47156 am+	**F**	FL		CD (U)	
47157 am	**FL**	P	*FL*	CD	Johnson Stevens Agencies
47186 am+	**RD**	E	*EW*	TI	Catcliffe Demon
47187	**F**	P		BL (U)	
47188 am+	**RD**	E		TI (U)	
47193	**F**	FL	*FL*	CD	
47194 am+	**F**	E	*EW*	TI	
47197	**FL**	P	*FL*	CD	
47200 am+	**RD**	E	*EW*	TI	Herbert Austin
47201 am+	**RD**	E	*EW*	TI	
47204 am+	**FL**	FL	*FL*	CD	
47205 am+	**FL**	FL	*FL*	CD	
47206	**FL**	P	*FL*	CD	The Morris Dancer
47207	**F**	FL	*FL*	CD	
47209 am+	**FL**	FL	*FL*	CD	
47210 am+	**F**	E	*EW*	TI	
47211 am+	**F**	E	*EW*	TI	
47212 +	**FL**	P	*FL*	CD	
47213 am+	**F**	E	*EW*	TI	Marchwood Military Port
47217 am+	**RD**	E	*EW*	TI	
47218 am+	**RD**	E	*EW*	TI	United Transport Europe
47219 am+	**RD**	E	*EW*	TI	Arnold Kunzler
47221 +	**F**	E		IM (U)	
47222 am+	**F**	E		TI (U)	
47223 +	**F**	E		CD (U)	
47224 +	**F**	E		IM (U)	
47225	**FL**	P	*FL*	CD	
47226 am+	**F**	E	*EW*	TI	
47228 am+	**RD**	E	*EW*	TI	axial
47229 am+	**RD**	E	*EW*	TI	
47231	**FL**	P	*FL*	CD	
47234 am+	**FL**	FL	*FL*	CD	
47236 am+	**RD**	E	*EW*	TI	ROVER GROUP QUALITY ASSURED

47237	am+	**RD**	E	*EW*	TI	
47238		**F**	E		BS (U)	
47241	am+	**RD**	E	*EW*	TI	Halewood Silver Jubilee 1988
47245	am+	**RD**	E	*EW*	TI	The Institute of Export
47256		**F**	E		DR (U)	
47258	am+	**RD**	FL	*FL*	CD	
47270		**FL**	P	*FL*	CD	Cory Brothers 1842–1992
47276	am+	**F**	E	*EW*	TI	
47277		**F**	E		IM (U)	
47278		**F**	E		SP (U)	
47279	am+	**FL**	P	*FL*	CD	
47280	am+	**F**	E	*EW*	TI	Pedigree
47281	am+	**F**	E	*EW*	TI	
47283		**FL**	P		CD (U)	
47284	am+	**F**	E	*EW*	TI	
47285	am+	**RD**	E	*EW*	TI	
47286	am+	**RD**	E	*EW*	TI	Port of Liverpool
47287	am+	**RD**	FL	*FL*	CD	
47289	am	**FL**	P	*FL*	CD	
47290	am+	**FL**	FL	*FL*	CD	
47291	am+	**F**	E		TI (U)	
47292	am+	**F**	FL	*FL*	CD	
47293	am+	**RD**	E	*EW*	TI	TRANSFESA
47294	+	**F**	E		TO (U)	
47295	+	**F**	FL	*FL*	CD	
47296		**FL**	P	*FL*	CD	
47297	am+	**RD**	E	*EW*	TI	Cobra RAILFREIGHT
47298	am+	**F**	E		TI (U)	Pegasus
47299	am+	**RD**	E		TI (U)	

Class 47/3. Built without Train Heat. (except 47300). RA6. Max. Speed 75 m.p.h.
All equipped with slow speed control.

47300		**C**	E		BS (U)	
47301		**FL**	P	*FL*	CD	Freightliner Birmingham
47302	a	**FR**	FL		TI (U)	
47303	am+	**FL**	FL	*FL*	CD	Freightliner Cleveland
47304	am+	**F**	E	*EW*	TI	
47305		**FL**	P	*FL*	CD	
47306	am+	**RD**	E	*EW*	TI	The Sapper
47307	am+	**RD**	E	*EW*	TI	
47308		**C**	E		BS (U)	
47309	am+	**F**	FL	*FL*	CD	The Halewood Transmission
47310	am+	**RD**	E	*EW*	TI	Henry Ford
47312	am+	**RD**	E	*EW*	TI	Parsec of Europe
47313	am+	**F**	E	*EW*	TI	
47314	am+	**F**	E	*EW*	TI	Transmark
47315		**C**	E	*EW*	IM	
47316	am+	**RD**	E	*EW*	TI	
47317		**F**	P		CD (U)	
47319	+	**F**	E		IM (U)	Norsk Hydro
47322		**FR**	P		CD (U)	

PLATFORM 5 PUBLISHING LIMITED
MAIL ORDER LIST

NEW TITLES

BR Pocket Book No.1: Locomotives £2.60
BR Pocket Book No.2: Coaching Stock £2.60
BR Pocket Book No.3: DMUs & Light Rail Systems £2.60
BR Pocket Book No.4: Electric Multiple Units £2.60
British Railways Locomotives & Coaching Stock 1998 **MARCH** £10.50
Light Rail Review 8 **MARCH** £9.50
British Railways Locomotives - The First 12 Years (SCTP) £18.95
London Underground Rolling Stock (Capital) £9.95
Railways around Lake Luzern (Bairstow) £9.95
London Tilbury & Southend Railway Part 2 (Kay) £9.95
Johnson's Atlas & Gazetteer of the Railways of Ireland (Midland) £19.99
The Londonderry & Lough Swilly Railway (Midland) £8.99
The Cavan & Leitrim Railway (Midland) £8.99
The 1998 Cowie Bus Handbook (British Bus) £15.00
The Fire Brigade Handbook: Special Appliances Vol. 2 (British Bus) £12.50

MODERN BRITISH RAILWAY TITLES

Preserved Locomotives of British Railways 9th ed. £7.95
Preserved Coaching Stock Part 1: BR Design Stock £7.95
Preserved Coaching Stock Part 2: Pre-Nationalisation Stock £8.95
Diesel & Electric Loco Register 3rd edition £7.95
Valley Lines - The People's Railway £9.95
Air Braked Series Wagon Fleet (SCTP) £7.95
Departmental Coaching Stock 5th edition (SCTP) £6.95
On-Track Plant on British Railways 5th edition (SCTP) £7.95
Engineers Series Wagon Fleet 970000-999999 (SCTP) £6.95
British Rail Wagon Fleet - B-Prefix Series (SCTP) £6.95
British Rail Internal Users (SCTP) £7.95
Private Owner Wagons Volume 1 (Metro) £7.95
Miles & Chains Volume 2 - London Midland (Milepost) £1.95
Miles & Chains Volume 3 - Scottish (Milepost) £1.95
Miles & Chains Volume 5 - Southern (Milepost) £1.95

OVERSEAS RAILWAYS

High Speed in Europe £9.95
High Speed in Japan £16.95

European Handbook No. 1: Benelux Railways 3rd edition £10.50
European Handbook No. 3: Austrian Railways 3rd edition £10.50
European Handbook No. 5: Swiss Railways 2nd edition £13.50
European Handbook No. 6: Italian Railways 1st edition £13.50
European Handbook No. 7: Irish Railways 1st edition £9.95
The Railways of Greece (Simms) £8.10
The Railways of Corsica (Simms) £5.10
Railways in the Austrian Tirol (Bairstow) £8.95
Irish Railways In Colour: From Steam to Diesel 1955-1967 (Midland) £16.99
Irish Railways In Colour: A Second Glance 1947-1970 (Midland) £19.99
Locomotives & Railcars of Bord Na Mona (Midland) £4.99

METRO SYSTEMS
World Metro Systems 2nd ed. (Capital) £10.95
The Twopenny Tube (Capital) .... [History of the Central Line] £5.95
Circles Under the Clyde (Capital) .... [Glasgow Subway] £15.95
Underground Official Handbook (Capital) £7.95
Docklands Light Rail Official Handbook (Capital) £7.95
Paris Metro Handbook (Capital) £7.95
The Berlin S-Bahn (Capital) £7.50
The Berlin U-Bahn (Capital) £7.50
Underground Architecture (Capital) £25.00
Mr Beck's Underground Map (Capital) £10.95

LIGHT RAIL TRANSIT AND TRAMS
Tram to Supertram .... [Sheffield Trams] £4.95
Light Rail Review 3 .... £7.50
Light Rail Review 4 .... £7.50
Light Rail Review 5 .... £7.50
Light Rail Review 6 .... £7.50
Light Rail Review 7 .... £8.95
Manx Electric £8.95
Light Rail in Europe (Capital) £9.95
London Tramways (Capital) £19.95
Tramway & Light Railway Atlas Germany 1996 (Blickpunkt Strassenbahn/LRTA) £10.45
The Tramways of Portugal (LRTA) £9.05

ATLASES, MAPS AND TRACK DIAGRAMS
Railway Track Diagrams No. 1: Scotland & Isle of Man (Quail) £6.50
British Railway Track Diagrams No. 4: Midland - 1990 Reprint (Quail) £6.95
Railway Track Diagrams No. 6: Ireland (Quail) £5.50

London Transport Railway Track Map (Quail) £1.75
Czech Republic & Slovakia Railway Map (Quail) £1.70
Berlin Track Map (Quail) £2.20
Moscow Railway Map (Quail) £2.20
Portugal Railway Map (Quail) £2.00
Greece Railway Map (Quail) £1.70
Poland Railway Map (Quail) £2.00
European Railway Atlas: France, Benelux (Ian Allan) £10.99
Track Diagram - South Yorkshire Supertram (HRT Rail Sales) £1.50
Track Diagram - Blackpool & Fleetwood (HRT Rail Sales) £1.00
Track Diagram - Tyne & Wear (HRT Rail Sales) £2.00

HISTORICAL RAILWAY TITLES
Steam Days on BR 1 - The Midland Line in Sheffield £4.95
Rails along the Sea Wall .... [Dawlish-Teignmouth Pictorial] £4.95
The Rolling Rivers £6.95
British Baltic Tanks £6.95
London Tilbury & Southend Railway Part 1 (Kay) £9.95
Signalling Atlas and Signal Box Directory Great Britain & Ireland (Kay) £9.95
Midland Railway System Maps: Leicester-London (Kay) £8.95
Mechanical Railway Signalling Part 1 (Kay) £9.50
Rails in the Isle of Wight (Midland) £16.99

ROAD TRANSPORT
London Coach Handbook (Capital) £15.00
London's Utility Buses (Capital) £19.95
London's Wartime Gas Buses (Capital) £5.95
Truckin' Round Scotland (Arthur Southern) £10.95
Bus Review 12 (Bus Enthusiast) £7.50
The Yorkshire Bus Handbook (British Bus) £12.50
The Ireland & Islands Bus Handbook (British Bus) £9.95
The Fire Brigade Handbook: Special Appliances Vol. 1 (British Bus) £12.50

RAMBLING
Rambles by Rail 2 - Liskeard-Looe £1.95
Rambles by Rail 4 - The New Forest £1.95
Buxton Spa Line Rail Rambles £1.20

POSTCARDS
Sheffield Supertram - Car No. 12 crosses a bridge over Sheffield Canal £0.30
Manchester Metrolink - Car No. 1021 in Aytoun Street £0.30

Quantity	Title		Price	Total
		SUB-TOTAL		
	Postage & Packing (see below for details)			
		TOTAL REMITTANCE		

Name: ....

Address: ....

....

.... Postcode: ....

Telephone No.: .... (Home) .... (Work)

Payment (Delete as appropriate)

I enclose my cheque (drawn on a UK bank) / British postal order for £ .... payable to:

'PLATFORM 5 PUBLISHING LTD'

Please debit my Visa / MasterCard / Access / Delta / Eurocard credit card for £ ....

Card No: .... Card Expiry Date: ....

Signature: .... Date: ....

Minimum credit card order accepted - £3.00.

Please send your remittance to:

Platform 5 Mail Order Department (PB)
3 Wyvern House, Sark Road
SHEFFIELD, S2 4HG, ENGLAND

If paying by credit card we can accept payment by post, or by telephone: UK - 0114 255 2625, Overseas - +44 114 255 2625, or fax: UK - 0114 255 2471, Overseas - +44 114 255 2471.

Postage & packing: please add: 10% UK (2nd Class); 20% Europe (Airmail); 30% Rest of World (Airfreight); 50% Rest of World (Airmail). If p&p works out at less than 40p, then please send 40p, this is the minimum post & packing accepted.

Please note that we cannot accept foreign currency cheques.

NOTE. When ordering publications in conjunction with a **Today's Railways** subscription offer please add on post & packing **before** deducting the voucher. Vouchers may **not** be combined.

Details correct as at 31st January 1998. Prices are not guaranteed and we reserve the right to alter details without further notification. Please allow 28 days for delivery in the UK.

47323	am+	**FL**	FL	*FL*	CD	
47326	am+	**RD**	E	*EW*	TI	Saltley Depot Quality Approved
47328	am+	**F**	E		TI (U)	
47329		**C**	FL		CD	
47330	am+	**F**	FL	*FL*	CD	
47331		**C**	E	*EW*	IM	
47332		**C**	FL	*FL*	CD	
47333		**C**	E		TO (U)	
47334	a	**FL**	FL	*FL*	CD	P&O Nedlloyd
47335	am+	**F**	E	*EW*	TI	
47337	am+	**FL**	P	*FL*	CD	
47338	am+	**RD**	E	*EW*	TI	
47339		**FL**	P	*FL*	CD	
47340		**C**	FL		ZC (U)	
47341		**C**	E		TO (U)	
47344	am+	**RD**	E	*EW*	TI	
47345		**FL**	P	*FL*	CD	
47347	a	**F**	P		CD (U)	
47348	am	**RD**	E	*EW*	TI	St. Christopher's Railway Home
47349		**FL**	P	*FL*	CD	
47350		**FO**	FL		CD (U)	
47351	am+	**RD**	E		TI (U)	
47352		**C**	E		FH (U)	
47353		**FL**	FL	*FL*	CD	
47354	a	**FL**	P	*FL*	CD	
47355	am+	**F**	E	*EW*	TI	
47356		**FO**	FL		BL (U)	
47357		**C**	E		BS (U)	
47358	am	**FL**	P	*FL*	CD	
47360	am+	**RD**	E	*EW*	TI	
47361	am+	**FL**	FL	*FL*	CD	
47362	am+	**F**	E	*EW*	TI	
47363	am+	**F**	E	*EW*	TI	
47365	am+	**RD**	E	*EW*	TI	ICI Diamond Jubilee
47366		**C**	E		SP (U)	
47367		**FL**	FL		TO (U)	
47368		**F**	E		SF (U)	
47369		**F**	E		IM (U)	
47370	am	**FL**	FL	*FL*	CD	Andrew A Hodgkinson
47371		**FL**	P	*FL*	CD	
47372		**C**	FL	*FL*	CD	
47375	am+	**RD**	E	*EW*	TI	Tinsley Traction Depot Quality Approved
47376		**FL**	P	*FL*	CD	Freightliner 1995
47377	a	**FL**	P	*FL*	CD	
47378	am+	**F**	E		TI (U)	
47379	am+	**F**	E	*EW*	TI	

Class 47/4. Equipped with train heating. RA6. Max. Speed 95 m.p.h.

| 47462 | | **R** | E | | TO (U) | |
| 47467 | | **BR** | E | *EW* | CD | |

47471		IO	E		CD (U)	
47473		BR	FL		ZC (U)	
47474		R	E	EW	IM	Sir Rowland Hill
47475		RX	E	EW	IM	Restive
47476		R	E	EW	IM	Night Mail
47478			E		BS (U)	
47481		BR	E		CD (U)	
47484		G	E		OC (U)	ISAMBARD KINGDOM BRUNEL
47488		W	FG		TM (U)	DAVIES THE OCEAN
47489		R	E		BS (U)	
47492		RX	E	EW	IM	
47501		R	E	EW	CD	
47513		BR	E		CD (U)	Severn
47519	+	G	E	EW	IM	
47520		I	E	EW	IM	
47522		R	E	EW	IM	
47523		M	E		TO (U)	
47524		RX	E		CD (U)	
47525		RD	FL		CD (U)	
47526		BR	E	EW	IM	
47528		M	E	EW	IM	The Queen's Own Mercian Yeomanry
47530		RX	E		CD (U)	
47532		RX	E		CD (U)	
47535		RX	E	EW	IM	
47536		RX	E		CD (U)	
47539		RX	E		CD (U)	
47540		C	FL		CD (U)	The Institution of Civil Engineers
47543		R	E	EW	IM	
47547		N	E		CD (U)	
47550		M	E		IM (U)	
47555		RD	E		TI (U)	The Commonwealth Spirit
47565		RX	E	EW	CD	Responsive
47566		RX	E		CD (U)	
47572		R	E	EW	CD	Ely Cathedral
47574		R	E		IM (U)	
47575		R	E	EW	CD	City of Hereford
47576		RX	E		CD (U)	
47584		RX	E	EW	CD	THE LOCOMOTIVE & CARRIAGE INSTITUTION 1911
47596		RX	E	EW	CD	
47624		RX	E	EW	CD	Saint Andrew
47627		0	E	EW	CD	
47628	j	RX	E		CD (U)	
47634		R	E	EW	CD	Holbeck
47635	j	R	E	EW	CD	
47640	j	R	E	EW	CD	University of Strathclyde

Class 47/6. Fitted with high phosphorus brake blocks. RA6. Max. Speed 75 m.p.h.

| 47676 | | I | E | | DR (U) | |

Class 47/7. Fitted with an older form of TDM. RA6. Max. Speed 95 m.p.h.
All have twin fuel tanks.

47701	**FG**	FG	*VX*	TM	Waverley
47702	**F**	E	*EW*	IM	County of Suffolk
47703	**FG**	FG	*VX*	TM	
47704	**RX**	E		CD (U)	
47705	**W**	LN	*SS*	CQ	GUY FAWKES
47709	**RX**	FG		TM (U)	
47710	**W**	FG		TM (U)	
47711	**N**	E	*EW*	IM	County of Hertfordshire
47712	**W**	FG	*VX*	TM	
47715	**N**	E		CD (U)	
47716	**RX**	E		CD (U)	
47717	**R**	E		CD (U)	

Class 47/7. Railnet dedicated locos. RA6. Max. Speed 95 m.p.h.
All have twin fuel tanks and are fitted with RCH jumper cables for operating
with propelling control vehicles (PCVs).

47721		**RX**	E	*EW*	CD	Saint Bede
47722	a	**RX**	E	*EW*	CD	The Queen Mother
47725		**RX**	E	*EW*	CD	The Railway Mission
47726		**RX**	E	*EW*	CD	Progress
47727	a	**RX**	E	*EW*	CD	Duke of Edinburgh's Award
47732		**RX**	E	*EW*	CD	Restormel
47733	a	**RX**	E	*EW*	CD	Eastern Star
47734		**RX**	E	*EW*	CD	Crewe Diesel Depot Quality Approved
47736	a	**RX**	E	*EW*	CD	Cambridge Traction
						& Rolling Stock Depot
47737		**RX**	E	*EW*	CD	Resurgent
47738	a	**RX**	E	*EW*	CD	Bristol Barton Hill
47739	a	**RX**	E	*EW*	CD	Resourceful
47741		**RX**	E	*EW*	CD	Resilient
47742		**RX**	E	*EW*	CD	The Enterprising Scot
47744	a	**E**	E	*EW*	CD	The Cornish Experience
47745		**RX**	E	*EW*	CD	Royal London Society for the Blind
47746	a	**RX**	E	*EW*	CD	The Bobby
47747	a	**RX**	E	*EW*	CD	Res Publica
47749		**RX**	E	*EW*	CD	Atlantic College
47750	a	**RX**	E	*EW*	CD	Royal Mail Cheltenham
47756	a	**RX**	E	*EW*	ML	Royal Mail Tyneside
47757	a	**RX**	E	*EW*	CD	Restitution
47758		**RX**	E	*EW*	CD	
47759		**RX**	E	*EW*	CD	
47760		**RX**	E	*EW*	CD	Restless
47761		**RX**	E	*EW*	CD	
47762		**RX**	E	*EW*	CD	
47763		**RX**	E	*EW*	CD	
47764		**RX**	E	*EW*	CD	Resounding
47765		**RX**	E	*EW*	CD	Ressaldar
47766		**RX**	E	*EW*	CD	Resolute

47767	a	**RX**	E	_EW_	ML	Saint Columba
47768		**RX**	E	_EW_	CD	Resonant
47769		**RX**	E	_EW_	CD	Resolve
47770		**RX**	E	_EW_	CD	Reserved
47771		**RX**	E	_EW_	CD	Heaton Traincare Depot
47772		**RX**	E	_EW_	CD	
47773	a	**RX**	E	_EW_	ML	Reservist
47774		**RX**	E	_EW_	CD	Poste Restante
47775		**RX**	E	_EW_	CD	Respite
47776		**RX**	E	_EW_	CD	Respected
47777		**RX**	E	_EW_	CD	Restored
47778		**RX**	E	_EW_	CD	Irresistible
47779		**RX**	E	_EW_	CD	
47780		**RX**	E	_EW_	CD	
47781		**RX**	E	_EW_	CD	Isle of Iona
47782		**RX**	E	_EW_	CD	
47783		**RX**	E	_EW_	CD	Saint Peter
47784		**RX**	E	_EW_	CD	Condover Hall
47785		**E**	E	_EW_	CD	Fiona Castle
47786	a	**E**	E	_EW_	CD	Roy Castle OBE
47787		**RX**	E	_EW_	CD	Victim Support
47788	a	**RX**	E	_EW_	CD	Captain Peter Manisty RN
47789	a	**RX**	E	_EW_	CD	Lindisfarne
47790	a	**RX**	E	_EW_	ML	Saint David/Dewi Sant
47791	a	**RX**	E	_EW_	CD	VENICE SIMPLON ORIENT EXPRESS
47792		**RX**	E	_EW_	CD	Saint Cuthbert
47793		**RX**	E	_EW_	CD	Saint Augustine

Class 47/4 continued. RA6. Max. Speed 95 m.p.h.

47798	a	**0**	E	_EW_	CD	Prince William
47799	a	**0**	E	_EW_	CD	Prince Henry
47802	+	**I**	E	_EW_	IM	
47803	+	**0**	E		SF (U)	
47805	a+	**I**	P	_VX_	CD	
47806	a+	**V**	P	_VX_	CD	
47807	a+	**PL**	P	_VX_	CD	
47810	a+	**I**	P	_VX_	CD	PORTERBROOK
47811	a+	**I**	P	_GW_	LA	
47812	a+	**I**	P	_VX_	CD	
47813	a+	**I**	P	_GW_	LA	
47814	a+	**V**	P	_VX_	CD	Totnes Castle
47815	a+	**I**	P	_GW_	LA	
47816	a+	**I**	P	_GW_	LA	Bristol Bath Road Quality Approved
47817	a+	**PL**	P	_VX_	CD	
47818	a+	**I**	P	_VX_	CD	
47822	a+	**V**	P	_VX_	CD	
47825	a+	**I**	P	_VX_	CD	Thomas Telford
47826	a+	**I**	P	_VX_	CD	
47827	a+	**I**	P	_VX_	CD	
47828	a+	**I**	P	_VX_	CD	
47829	a+	**I**	P	_VX_	CD	

47830	a+	I	P		ZC (U)	
47831	a+	I	P	VX	CD	Bolton Wanderer
47832	a+	I	P	GW	LA	
47839	a+	I	P	VX	CD	
47840	a+	I	P	VX	CD	NORTH STAR
47841	a+	I	P	VX	CD	The Institution of
						Mechanical Engineers
47843	a+	I	P	VX	CD	
47844	a+	V	P	VX	CD	
47845	a+	V	P	VX	CD	County of Kent
47846	a+	0	P	GW	LA	THOR
47847	a+	I	P	VX	CD	
47848	a+	I	P	VX	CD	
47849	a+	I	P	VX	CD	
47851	a+	I	P	VX	CD	
47853	a+	I	P	VX	CD	
47854	a+	I	P	VX	CD	Women's Royal Voluntary Service
47971	*	BR	E	EW	CD	Robin Hood
47972		CS	E	EW	IM	The Royal Army Ordnance Corps
47976	*	C	E	EW	CD	Aviemore Centre

Class 47/3 continued. RA6. Max. Speed 75 m.p.h.

47981		C	E	EW	IM	

CLASS 50   ENGLISH ELECTRIC TYPE 4   Co–Co

Built: 1967–68 by English Electric at Vulcan Foundry, Newton-le-Willows.
Engine: English Electric 16CVST of 2010 kW (2700 hp) at 850 r.p.m.
Main Generator: English Electric 840/4B.
Traction Motors: English Electric 538/5A.
Max. Tractive Effort: 216 kN (48500 lbf).
Cont. Tractive Effort: 147 kN (33000 lbf) at 23.5 m.p.h.
Power At Rail: 1540 kW (2070 hp).   **Length over Buffers:** 20.88 m.
Brake Force: 59 t.   **Wheel Diameter:** 1092 mm.
Design Speed: 105 m.p.h.   **Weight:** 117 t.
Max. Speed: 90 m.p.h.   **RA:** 6.
Train Brakes: Air & Vacuum.
Multiple Working: Orange Square Coupling Code (within class only).
ETH Index: 66.
All equipped with slow speed control.

50031		BR	FF	SS	KR	Hood

CLASS 55   DELTIC   Co–Co

Built: 1961 by English Electric at Vulcan Foundry, Newton-le-Willows.
Engine: Two Napier-Deltic T18-25 of 1230 kW (1650 h.p.) at 1500 r.p.m.
Main Generators: Two English Electric EE829.
Traction Motors: EE538 axle-hung.
Max. Tractive Effort: 222 kN (50000 lbf).
Cont. Tractive Effort: 136 kN (30500 lbf) at 32.5 m.p.h.
Power At Rail: 1969 kW (2640 hp).   **Length over Buffers:** 17.65 m.
Brake Force: 51 t.   **Wheel Diameter:** 1092 mm.
Design Speed: 100 m.p.h.   **Weight:** 105 t.
Max. Speed: 100 m.p.h.   **RA:** 5.
Train Brakes: Air & Vacuum.   **Multiple Working:** Not equipped.
ETH Index: 66.

55022	**G**	NT	*SS*	BN	ROYAL SCOTS GREY

CLASS 56   BRUSH TYPE 5   Co–Co

Built: 1976–84 by Electroputere at Craiova, Romania (as sub contractors for Brush) or BREL at Doncaster or Crewe Works.
Engine: Ruston Paxman 16RK3CT of 2460 kW (3250 hp) at 900 rpm.
Main Alternator: Brush BA1101A.
Traction Motors: Brush TM73-62.
Max. Tractive Effort: 275 kN (61800 lbf).
Cont. Tractive Effort: 240 kN (53950 lbf) at 16.8 m.p.h.
Power At Rail: 1790 kW (2400 hp).   **Length over Buffers:** 19.36 m.
Brake Force: 60 t.   **Wheel Diameter:** 1143 mm.
Design Speed: 80 m.p.h.   **Weight:** 125 t.
Max. Speed: 80 m.p.h.   **RA:** 7.
Train Brakes: Air.
Multiple Working: Red Diamond Coupling Code.
All equipped with slow speed control.

56003	**LH**	E	*EW*	IM	
56004		E	*EW*	IM	
56006	**LH**	E	*EW*	IM	Ferrybridge 'C' Power Station
56007	**FT**	E	*EW*	IM	
56008		E		IM (U)	
56010	**FT**	E	*EW*	IM	
56011	**F**	E	*EW*	IM	
56012	**F**	E		IM (U)	
56014	**F**	E		IM (U)	
56018	**E**	E	*EW*	IM	
56019	**FR**	E	*EW*	IM	
56021	**LH**	E	*EW*	IM	
56022	**FT**	E	*EW*	IM	
56025	**FT**	E	*EW*	IM	
56027	**LH**	E	*EW*	IM	
56029	**F**	E	*EW*	IM	
56031	**C**	E	*EW*	IM	

56032	E	E	*EW*	IM	
56033	FT	E	*EW*	IM	Shotton Paper Mill
56034	LH	E	*EW*	IM	Castell Ogwr/Ogmore Castle
56035	LH	E	*EW*	IM	
56036	CT	E	*EW*	IM	
56037	E	E	*EW*	IM	
56038	FT	E	*EW*	IM	Western Mail
56039	LH	E	*EW*	IM	
56040	FT	E	*EW*	IM	Oystermouth
56041	E	E	*EW*	IM	
56043	F	E	*EW*	IM	
56044	FT	E	*EW*	IM	Cardiff Canton Quality Assured
56045	LH	E	*EW*	IM	British Steel Shelton
56046	C	E	*EW*	IM	
56047	CT	E	*EW*	IM	
56048	C	E	*EW*	IM	
56049	CT	E	*EW*	IM	
56050	LH	E	*EW*	IM	British Steel Teeside
56051	E	E	*EW*	IM	
56052	FT	E	*EW*	IM	The Cardiff Rod Mill
56053	FT	E	*EW*	IM	Sir Morgannwg Ganol/
					County of Mid Glamorgan
56054	FT	E	*EW*	IM	British Steel Llanwern
56055	LH	E	*EW*	IM	
56056	FT	E	*EW*	IM	
56057	E	E	*EW*	IM	British Fuels
56058	E	E	*EW*	IM	
56059	E	E	*EW*	IM	
56060	E	E	*EW*	IM	
56061	F	E	*EW*	IM	
56062	F	E	*EW*	IM	Mountsorrel
56063	F	E	*EW*	IM	Bardon Hill
56064	FT	E	*EW*	IM	
56065	E	E	*EW*	IM	
56066	FT	E	*EW*	IM	
56067	E	E	*EW*	IM	
56068	E	E	*EW*	IM	
56069	F	E	*EW*	IM	Thornaby TMD
56070	FT	E	*EW*	IM	
56071	FT	E	*EW*	IM	
56072	FT	E	*EW*	IM	
56073	FT	E	*EW*	IM	
56074	LH	E	*EW*	IM	Kellingley Colliery
56075	F	E	*EW*	IM	West Yorkshire Enterprise
56076	F	E	*EW*	IM	
56077	LH	E	*EW*	IM	Thorpe Marsh Power Station
56078	F	E	*EW*	IM	
56079	FT	E	*EW*	IM	
56080	F	E	*EW*	IM	Selby Coalfield
56081	F	E	*EW*	IM	
56082	F	E	*EW*	IM	

56083	LH	E	EW	IM	
56084	LH	E	EW	IM	
56085	LH	E	EW	IM	
56086	FT	E	EW	IM	The Magistrates' Association
56087	E	E	EW	IM	ABP Port of Hull
56088	E	E	EW	IM	
56089	E	E	EW	IM	
56090	LH	E	EW	IM	
56091	F	E	EW	IM	Castle Donington Power Station
56092	FT	E	EW	IM	
56093	FT	E	EW	IM	The Institution of Mining Engineers
56094	F	E	EW	IM	Eggborough Power Station
56095	F	E	EW	IM	Harworth Colliery
56096	E	E	EW	IM	
56097	F	E	EW	IM	
56098	F	E	EW	IM	
56099	FT	E	EW	IM	Fiddlers Ferry Power Station
56100	LH	E	EW	IM	
56101	FT	E	EW	IM	Mutual Improvement
56102	LH	E	EW	IM	
56103	E	E	EW	IM	Stora
56104	F	E	EW	IM	
56105	E	E	EW	IM	
56106	LH	E	EW	IM	
56107	LH	E	EW	IM	
56108	F	E	EW	IM	
56109	LH	E	EW	IM	
56110	LH	E	EW	IM	Croft
56111	LH	E	EW	IM	
56112	LH	E	EW	IM	Stainless Pioneer
56113	FT	E	EW	IM	
56114	E	E	EW	IM	Maltby Colliery
56115	FT	E	EW	IM	
56116	LH	E	EW	IM	
56117	E	E	EW	IM	
56118	LH	E	EW	IM	
56119	E	E	EW	IM	
56120	E	E	EW	IM	
56121	F	E	EW	IM	
56123	FT	E	EW	IM	Drax Power Station
56124	F	E	EW	IM	
56125	FT	E	EW	IM	
56126	F	E	EW	IM	
56127	FT	E	EW	IM	
56128	FT	E	EW	IM	
56129	FT	E	EW	IM	
56130	LH	E	EW	IM	Wardley Opencast
56131	F	E	EW	IM	Ellington Colliery
56132	FT	E	EW	IM	
56133	FT	E	EW	IM	Crewe Locomotive Works
56134	F	E	EW	IM	Blyth Power

56135   F   E   *EW*   IM   Port of Tyne Authority

CLASS 58   BREL TYPE 5   Co–Co

Built: 1983–87 by BREL at Doncaster Works.
Engine: Ruston Paxman RK3ACT of 2460 kW (3300 hp) at 1000 rpm.
Main Alternator: Brush BA1101B.
Traction Motors: Brush TM73-62.
Max. Tractive Effort: 275 kN (61800 lbf).
Cont. Tractive Effort: 240 kN (53950 lbf) at 17.4 m.p.h.
Power At Rail: 1780 kW (2387 hp).   **Length over Buffers:** 19.13 m.
Brake Force: 62 t.   **Wheel Diameter:** 1120 mm.
Design Speed: 80 m.p.h.   **Weight:** 130 t.
Max. Speed: 80 m.p.h.   **RA:** 7.
Train Brakes: Air.
Multiple Working: Red Diamond Coupling Code.
All equipped with slow speed control.

58001	**FM**	E	*EW*	TO	
58002	**ML**	E	*EW*	TO	Daw Mill Colliery
58003	**FM**	E	*EW*	TO	Markham Colliery
58004	**FM**	E	*EW*	TO	
58005	**ML**	E	*EW*	TO	Ironbridge Power Station
58006	**F**	E	*EW*	TO	
58007	**FM**	E	*EW*	TO	Drakelow Power Station
58008	**ML**	E	*EW*	TO	
58009	**FM**	E	*EW*	TO	
58010	**FM**	E	*EW*	TO	
58011	**FM**	E	*EW*	TO	Worksop Depot
58012	**FM**	E	*EW*	TO	
58013	**ML**	E	*EW*	TO	
58014	**ML**	E	*EW*	TO	Didcot Power Station
58015	**FM**	E	*EW*	TO	
58016	**E**	E	*EW*	TO	
58017	**FM**	E	*EW*	TO	Eastleigh Depot
58018	**FM**	E	*EW*	TO	High Marnham Power Station
58019	**FM**	E	*EW*	TO	Shirebrook Colliery
58020	**FM**	E	*EW*	TO	Doncaster Works
58021	**ML**	E	*EW*	TO	Hither Green Depot
58022	**FM**	E	*EW*	TO	
58023	**ML**	E	*EW*	TO	Peterborough Depot
58024	**E**	E	*EW*	TO	
58025	**E**	E	*EW*	TO	
58026	**FM**	E	*EW*	TO	
58027	**FM**	E	*EW*	TO	
58028	**FM**	E	*EW*	TO	
58029	**FM**	E	*EW*	TO	
58030	**E**	E	*EW*	TO	
58031	**FM**	E	*EW*	TO	
58032	**ML**	E	*EW*	TO	Thoresby Colliery
58033	**E**	E	*EW*	TO	

58034	FM	E	*EW*	TO	Bassetlaw
58035	FM	E	*EW*	TO	
58036	ML	E	*EW*	TO	
58037	E	E	*EW*	TO	
58038	ML	E	*EW*	TO	
58039	E	E	*EW*	TO	
58040	FM	E	*EW*	TO	Cottam Power Station
58041	FM	E	*EW*	TO	Ratcliffe Power Station
58042	ML	E	*EW*	TO	Petrolea
58043	FM	E	*EW*	TO	
58044	FM	E	*EW*	TO	Oxcroft Opencast
58045	FM	E	*EW*	TO	
58046	ML	E	*EW*	TO	Asfordby Mine
58047	E	E	*EW*	TO	
58048	E	E	*EW*	TO	
58049	E	E	*EW*	TO	Littleton Colliery
58050	E	E	*EW*	TO	

CLASS 59   GENERAL MOTORS TYPE 5   Co–Co

Built: 1985 (59001–4), 1989 (59005) by General Motors, La Grange, Illinois, U.S.A. or 1990 (59101–4), 1994 (59201) and 1995 (59202–6) by General Motors, London, Ontario, Canada.
Engine: General Motors 645E3C two stroke of 2460 kW (3300 hp) at 900 rpm.
Main Alternator: General Motors AR11 MLD-D14A.
Traction Motors: General Motors D77B.
Max. Tractive Effort: 506 kN (113 550 lbf).
Cont. Tractive Effort: 291 kN (65 300 lbf) at 14.3 m.p.h.
Power At Rail: 1889 kW (2533 hp).   **Length over Buffers:** 21.35 m.
Brake Force: 69 t.   **Wheel Diameter:** 1067 mm.
Weight: 121 t.   **RA:** 7.
Design Speed: 60 m.p.h. (75 m.p.h. Class 59/2).
Max. Speed: 60 m.p.h. (75 m.p.h. Class 59/2).

Class 59/0. Owned by Foster-Yeoman Ltd.

59001	FY	FY	*MD*	MD	YEOMAN ENDEAVOUR
59002	FY	FY	*MD*	MD	ALAN J DAY
59004	FY	FY	*MD*	MD	PAUL A HAMMOND
59005	FY	FY	*MD*	MD	KENNETH J. PAINTER

Class 59/1. Owned by ARC Limited.

59101	AC	AC	*MD*	WH	Village of Whatley
59102	AC	AC	*MD*	WH	Village of Chantry
59103	AC	AC	*MD*	WH	Village of Mells
59104	AC	AC	*MD*	WH	Village of Great Elm

Class 59/2. Owned by National Power.

59201	NP	NP	*NP*	FB	Vale of York
59202	NP	NP	*NP*	FB	Vale of White Horse
59203	NP	NP	*NP*	FB	Vale of Pickering

59204	**NP**	NP	*NP*	FB	Vale of Glamorgan
59205	**NP**	NP	*NP*	FB	Vale of Evesham
59206	**NP**	NP	*NP*	FB	Pride of Ferrybridge

CLASS 60   BRUSH TYPE 5   Co–Co

Built: 1989–1993 by Brush Traction at Loughborough.
Engine: Mirrlees MB275T of 2310 kW (3100 hp) at 1000 rpm.
Main Alternator: Brush .
Traction Motors: Brush separately excited.
Max. Tractive Effort: 500 kN (106500 lbf).
Cont. Tractive Effort: 336 kN (71570 lbf) at 17.4 m.p.h.
Power At Rail: 1800 kW (2415 hp).   **Length over Buffers:** 21.34 m.
Brake Force: 74 t.   **Wheel Diameter:** 1118 mm.
Design Speed: 62 m.p.h.   **Weight:** 129 t.
Max. Speed: 60 m.p.h.   **RA:** 7.
Multiple Working: Within class only.
All equipped with slow speed control.

Non-standard liveries:
60006 & 60033 are British Steel Trafalgar blue.

60001		**E**	E	*EW*	TO	
60002	+	**E**	E	*EW*	TO	
60003	+	**E**	E	*EW*	TO	FREIGHT TRANSPORT ASSOCIATION
60004	+	**E**	E	*EW*	TO	
60005		**E**	E	*EW*	TO	
60006		**O**	E	*EW*	TO	Scunthorpe Ironmaster
60007	+	**LH**	E	*EW*	TO	
60008		**LH**	E	*EW*	TO	GYPSUM QUEEN II
60009	+	**E**	E	*EW*	TO	
60010		**E**	E	*EW*	TO	
60011		**ML**	E	*EW*	TO	
60012	+	**E**	E	*EW*	TO	
60013		**F**	E	*EW*	TO	Robert Boyle
60014		**E**	E	*EW*	TO	
60015	+	**FT**	E	*EW*	TO	Bow Fell
60016		**E**	E	*EW*	TO	
60017	+	**E**	E	*EW*	TO	Shotton Works Centenary Year 1996
60018		**E**	E	*EW*	TO	
60019		**E**	E	*EW*	TO	
60020	+	**E**	E	*EW*	TO	
60021	+	**F**	E	*EW*	TO	Pen-y-Ghent
60022	+	**E**	E	*EW*	TO	
60023	+	**E**	E	*EW*	TO	
60024		**E**	E	*EW*	TO	
60025	+	**LH**	E	*EW*	TO	
60026	+	**E**	E	*EW*	TO	
60027	+	**E**	E	*EW*	TO	
60028	+	**E**	E	*EW*	TO	
60029		**E**	E	*EW*	TO	

60030		E	E	EW	TO	
60031		F	E	EW	TO	
60032		FT	E	EW	TO	William Booth
60033		0	E	EW	TO	Tees Steel Express
60034		FT	E	EW	TO	Carnedd Llewelyn
60035		FT	E	EW	TO	Florence Nightingale
60036		E	E	EW	TO	
60037	+	E	E	EW	TO	Aberddawan/Aberthaw
60038	+	LH	E	EW	TO	
60039		E	E	EW	TO	
60040		E	E	EW	TO	
60041	+	E	E	EW	TO	
60042		E	E	EW	TO	
60043		E	E	EW	TO	
60044		ML	E	EW	TO	
60045		E	E	EW	TO	The Permanent Way Institution
60046		E	E	EW	TO	
60047	+	E	E	EW	TO	
60048		E	E	EW	TO	Eastern
60049		E	E	EW	TO	
60050		E	E	EW	TO	
60051	+	E	E	EW	TO	
60052	+	E	E	EW	TO	
60053		E	E	EW	TO	Nordic Terminal
60054	+	F	E	EW	TO	Charles Babbage
60055		FT	E	EW	TO	Thomas Barnardo
60056	+	FT	E	EW	TO	William Beveridge
60057		F	E	EW	TO	Adam Smith
60058		FT	E	EW	TO	John Howard
60059	+	LH	E	EW	TO	Swinden Dalesman
60060		F	E	EW	TO	James Watt
60061		FT	E	EW	TO	Alexander Graham Bell
60062		FT	E	EW	TO	Samuel Johnson
60063		FT	E	EW	TO	James Murray
60064	+	FH	E	EW	TO	Back Tor
60065		FT	E	EW	TO	Kinder Low
60066		FT	E	EW	TO	John Logie Baird
60067		F	E	EW	TO	James Clerk-Maxwell
60068		F	E	EW	TO	Charles Darwin
60069		F	E	EW	TO	Humphry Davy
60070	+	FH	E	EW	TO	John Loudon McAdam
60071	+	FM	E	EW	TO	Dorothy Garrod
60072		FM	E	EW	TO	
60073		FM	E	EW	TO	
60074		FM	E	EW	TO	Braeriach
60075		FM	E	EW	TO	
60076		FM	E	EW	TO	
60077	+	FM	E	EW	TO	
60078		ML	E	EW	TO	
60079		FM	E	EW	TO	Foinaven
60080	+	FT	E	EW	TO	Kinder Scout

60081	+	FT	E	*EW*	TO
60082		F	E	*EW*	TO	Mam Tor
60083		E	E	*EW*	TO
60084		FT	E	*EW*	TO	Cross Fell
60085		FT	E	*EW*	TO
60086		FM	E	*EW*	TO	Schiehallion
60087		FM	E	*EW*	TO	Slioch
60088		FM	E	*EW*	TO	Buachaille Etive Mor
60089		FT	E	*EW*	TO	Arcuil
60090	+	F	E	*EW*	TO	Quinag
60091		F	E	*EW*	TO	An Teallach
60092		FT	E	*EW*	TO	Reginald Munns
60093		FT	E	*EW*	TO	Jack Stirk
60094		FM	E	*EW*	TO	Tryfan
60095		F	E	*EW*	TO
60096	+	E	E	*EW*	TO	Ben Macdui
60097		FT	E	*EW*	TO	Pillar
60098	+	E	E	*EW*	TO	Charles Francis Brush
60099		FM	E	*EW*	TO	Ben More Assynt
60100		FM	E	*EW*	TO	Boar of Badenoch

CLASS 66   GENERAL MOTORS TYPE 5   Co–Co

Built: 1998 onwards by General Motors, London, Ontario, Canada.
Engine: General Motors 12N-710-3GB-EC two stroke of 2460 kW (3300 hp) at 900 rpm.
Main Alternator: General Motors AR8.
Traction Motors: General Motors D43TR.
Max. Tractive Effort: 399 kN (89800 lbf).
Cont. Tractive Effort: 253 kN (57000 lbf) at 15.8 m.p.h.

Power At Rail:	**Length over Buffers:** 21.39 m.
Brake Force:	**Wheel Diameter:** 1120 mm.
Weight:	**RA:** 7.
Design Speed: 75 m.p.h.	
Max. Speed: 75 m.p.h.	

66001
66002
66003
66004
66005
66006
66007
66008
66009
66010
66011
66012
66013
66014
66015
66016

66017
66018
66019
66020
66021
66022
66023
66024
66025
66026
66027
66028
66029
66030
66031
66032
66033
66034
66035
66036
66037
66038
66039
66040
66041
66042
66043
66044
66045
66046
66047
66048
66049
66050
66051
66052
66053
66054
66055
66056
66057
66058
66059
66060
66061
66062
66063
66064
66065
66066
66067

66068
66069
66070
66071
66072
66073
66074
66075
66076
66077
66078
66079
66080
66081
66082
66083
66084
66085
66086
66087
66088
66089
66090
66091
66092
66093
66094
66095
66096
66097
66098
66099
66100

On order to 66250.

2. ELECTRIC LOCOMOTIVES

CLASS 73/0   ELECTRO-DIESEL   Bo–Bo

Built: 1962 by BR at Eastleigh Works.
Supply System: 660–850 V d.c. from third rail.
Engine: English Electric 4SRKT of 447 kW (600 hp) at 850 rpm.
Main Generator: English Electric 824/3D.
Traction Motors: English Electric 542A.
Max. Tractive Effort: Electric 187 kN (42000 lbf). Diesel 152 kN (34100 lbf).
Continuous Rating: Electric 1060 kW (1420 hp) giving a tractive effort of 43 kN (9600 lbf) at 55.5 m.p.h.
Cont. Tractive Effort: Diesel 72 kN (16100 lbf) at 10 m.p.h.
Maximum Rail Power: Electric 1830 kW (2450 hp) at 37 m.p.h.

Brake Force: 31 t.	**Length over Buffers:** 16.36 m.
Design Speed: 80 m.p.h.	**Weight:** 76.5 t.
Max. Speed: 60 m.p.h.	**RA:** 6.
Wheel Diameter: 1016 mm.	**ETH Index (Elec. power):** 66.

Train Brakes: Air, Vacuum and electro-pneumatic.
Multiple Working: Within sub-class, with Class 33/1 and various 750 V d.c. EMUs.
Couplings: Drop-head buckeye.

Formerly numbered E 6002/5.

73002	**BR**	ME		BD (U)
73005		ME	*ME*	BD

CLASS 73/1 & 73/2   ELECTRO-DIESEL   Bo–Bo

Built: 1965–67 by English Electric Co. at Vulcan Foundry, Newton le Willows.
Supply System: 660–850 V d.c. from third rail.
Engine: English Electric 4SRKT of 447 kW (600 hp) at 850 rpm.
Main Generator: English Electric 824/5D.
Traction Motors: English Electric 546/1B.
Max. Tractive Effort: Electric 179 kN (40000 lbf). Diesel 160 kN (36000 lbf).
Continuous Rating: Electric 1060 kW (1420 hp) giving a tractive effort of 35 kN (7800 lbf) at 68 m.p.h.
Cont. Tractive Effort: Diesel 60 kN (13600 lbf) at 11.5 m.p.h.
Maximum Rail Power: Electric 2350 kW (3150 hp) at 42 m.p.h.

Brake Force: 31 t.	**Length over Buffers:** 16.36 m.
Design Speed: 90 m.p.h.	**Weight:** 77 t.
Max. Speed: 60 (90*) m.p.h.	**RA:** 6.
Wheel Diameter: 1016 mm.	**ETH Index (Elec. power):** 66.

Train Brakes: Air, Vacuum and electro-pneumatic.
Multiple Working: Within sub-class, with Class 33/1 and various 750 V d.c. EMUs.
Couplings: Drop-head buckeye.

Non-standard Livery:
73101 is Pullman umber & cream.

a Vacuum brake isolated.

Formerly numbered E 6001–20/22–26/28–49 (not in order).

73101		**0**	E	*EW*	HG	The Royal Alex'
73103		**I0**	E	*EW*	HG	
73104		**I0**	E	*EW*	HG	
73105		**C**	E	*EW*	HG	
73106		**D**	E	*EW*	HG	
73107		**C**	E	*EW*	HG	Redhill 1844–1994
73108		**C**	E	*EW*	HG	
73109	*	**ST**	SW	*SW*	BM	Battle of Britain 50th Anniversary
73110		**C**	E	*EW*	HG	
73114		**ML**	E	*EW*	HG	Stewarts Lane Traction Maintenance Depot
73117		**I0**	E	*EW*	HG	University of Surrey
73118	c	**EP**	LC	*ES*	OC	
73119		**C**	E	*EW*	HG	Kentish Mercury
73126		**N**	E		OC (U)	
73128		**E**	E	*EW*	HG	
73129		**N**	E	*EW*	HG	City of Winchester
73130	c	**EP**	LC	*ES*	OC	
73131		**E**	E	*EW*	HG	
73132		**I0**	E	*EW*	HG	
73133		**ML**	E	*EW*	HG	The Bluebell Railway
73134		**I0**	E	*EW*	HG	Woking Homes 1885–1985
73136		**ML**	E	*EW*	HG	Kent Youth Music
73138		**C**	E	*EW*	HG	
73139		**I0**	E	*EW*	HG	
73140		**I0**	E	*EW*	HG	
73141		**I0**	E	*EW*	HG	
73201	a*	**GX**	P	*GX*	SL	Broadlands
73202	a*	**GX**	P	*GX*	SL	Royal Observer Corps
73203	a*	**GX**	P	*GX*	SL	
73204	a*	**GX**	P	*GX*	SL	Stewarts Lane 1860–1985
73205	a*	**GX**	P	*GX*	SL	
73206	a*	**GX**	P	*GX*	SL	Gatwick Express
73207	a*	**GX**	P	*GX*	SL	County of East Sussex
73208	a*	**GX**	P	*GX*	SL	Croydon 1883–1983
73209	a*	**GX**	P	*GX*	SL	
73210	a*	**GX**	P	*GX*	SL	Selhurst
73211	a*	**GX**	P	*GX*	SL	
73212	a*	**GX**	P	*GX*	SL	Airtour Suisse
73213	a*	**GX**	P	*GX*	SL	University of Kent at Canterbury
73235	a*	**GX**	P	*GX*	SL	

CLASS 73/9   ELECTRO-DIESEL   Bo–Bo

For details see Class 73/0. Sandite fitted locos.

Formerly numbered E 6001/6.

73901	**MD**	ME	*ME*	BD
73906	**MD**	ME	*ME*	BD

NOTES FOR CLASSES 86–91.

The following common features apply to all locos of Classes 86–91.

Supply System: 25 kV a.c. from overhead equipment.
Communication Equipment: Driver–guard telephone.
Multiple Working: Time division multiplex system.

a vacuum brake isolated.

Class 86 were formerly numbered E 3101–3200 (not in order).

CLASS 86/1   BR DESIGN   Bo–Bo

Built: 1965–66 by English Electric Co. at Vulcan Foundry, Newton le Willows or BR at Doncaster Works. Rebuilt with Class 87 type bogies and motors. Tap changer control.
Traction Motors: GEC G412AZ frame mounted.
Max. Tractive Effort: 258 kN (58000 lbf).
Continuous Rating: 3730 kW (5000 hp) giving a tractive effort of 95 kN (21300 lbf) at 87 m.p.h.
Maximum Rail Power: 5860 kW (7860 hp) at 50.8 m.p.h.
Brake Force: 40 t.   **Length over Buffers**: 17.83 m.
Design Speed: 110 m.p.h.   **Weight**: 87 t.
Max. Speed: 110 m.p.h.   **RA**: 6.
ETH Index: 74.   **Wheel Diameter**: 1150 mm.
Train Brakes: Air & Vacuum.   **Electric Brake**: Rheostatic.

86101	I	F	ZC (U)	Sir William A Stanier FRS
86102	I	F	ZC (U)	Robert A Riddles
86103	I	F	ZC (U)	André Chapelon

CLASS 86/2   BR DESIGN   Bo–Bo

Built: 1965–66 by English Electric Co. at Vulcan Foundry, Newton le Willows or BR at Doncaster Works. Later rebuilt with resilient wheels and flexicoil suspension. Tap changer control.
Traction Motors: AEI 282BZ.
Max. Tractive Effort: 207 kN (46500 lbf).
Continuous Rating: 3010 kW (4040 hp) giving a tractive effort of 85 kN (19200 lbf) at 77.5 m.p.h.
Maximum Rail Power: 4550 kW (6100 hp) at 49.5 m.p.h.
Brake Force: 40 t.   **Length over Buffers**: 17.83 m.

Design Speed: 125 m.p.h. **Weight:** 85 t–86 t.
Max. Speed: 100 (110*) m.p.h. **RA:** 6.
ETH Index: 66 (75§). **Wheel Diameter:** 1156 mm.
Train Brakes: Air & Vacuum. **Electric Brake:** Rheostatic.

86204		I	F	VX	LG	City of Carlisle
86205	a	I	F	VX	LG	City of Lancaster
86206	a	I	F	VX	LG	City of Stoke on Trent
86207	a	I	F	VW	WN	City of Lichfield
86208		I	E	EW	CE	City of Chester
86209		I	F	VW	WN	City of Coventry
86210		RX	E	EW	CE	C.I.T. 75th Anniversary
86212	a	I	F	VX	LG	Preston Guild 1328–1992
86213		I	F	VX	LG	Lancashire Witch
86214	a	I	F	VX	LG	Sans Pareil
86215		I	F	AR	NC	
86216	a	I	F	VW	WN	Meteor
86217	a	I	F	AR	NC	City University
86218		I	F	AR	NC	YEAR OF OPERA & MUSICAL THEATRE 1997
86219	a	I	F		ZH (U)	Phoenix
86220	a§	I	F	AR	NC	The Round Tabler
86221		I	F	AR	NC	B.B.C. Look East
86222	a	I	F	VX	LG	Clothes Show Live
86223	a§	I	F	AR	NC	Norwich Union
86224	a	I	F	VW	WN	Caledonian
86225	a	I	F	VW	WN	Hardwicke
86226	a	I	F	VX	LG	CHARLES RENNIE MACKINTOSH
86227	a§	I	F	VX	LG	Sir Henry Johnson
86228		I	F	AR	NC	Vulcan Heritage
86229	a	I	F	VX	LG	Sir John Betjeman
86230		I	F	AR	NC	
86231	*	I	F	VW	WN	Starlight Express
86232	a	I	F	AR	NC	Norfolk and Norwich Festival
86233	a	I	F	VX	LG	Laurence Olivier
86234	a	I	F	VX	LG	J B Priestley OM
86235		I	F	AR	NC	Crown Point
86236	a	I	F	VW	WN	Josiah Wedgwood MASTER POTTER 1736–1795
86237		I	F	AR	NC	University of East Anglia
86238	a	I	F	AR	NC	European Community
86240		I	F	VW	WN	Bishop Eric Treacy
86241		RX	E	EW	CE	Glenfiddich
86242		I	F	VW	WN	James Kennedy GC
86243		RX	E	EW	CE	
86244	a	I	F	VX	LG	The Royal British Legion
86245	a	I	F	VW	WN	Dudley Castle
86246	a	I	F	AR	NC	Royal Anglian Regiment
86247	a	I	F	VX	LG	Abraham Darby
86248		I	F	VW	WN	Sir Clwyd/County of Clwyd
86249	a	I	F		ZC (U)	County of Merseyside

86250	a	I	F	AR	NC	The Glasgow Herald
86251		I	F	VW	WN	The Birmingham Post
86252	a	I	F	VX	LG	The Liverpool Daily Post
86253		I	F	VW	WN	The Manchester Guardian
86254		RX	E	EW	CE	
86255	a	I	F	VX	LG	Penrith Beacon
86256		I	F	VW	WN	Pebble Mill
86257	a	I	F	.	NC (U)	Snowdon
86258		I	F	VW	WN	Talyllyn—The First Preserved Railway
86259	a	I	F	VX	LG	Greater MANCHESTER
						THE LIFE & SOUL OF BRITAIN
86260	a	I	F	VX	LG	Driver Wallace Oakes G.C.
86261		E	E	EW	CE	THE RAIL CHARTER PARTNERSHIP

CLASS 86/4 & 86/6   BR DESIGN   Bo–Bo

Built: 1965–66 by English Electric Co. at Vulcan Foundry, Newton le Willows or BR at Doncaster Works. Later rebuilt with resilient wheels and flexicoil suspension. Tap changer control.
Traction Motors: AEI 282AZ.
Max. Tractive Effort: 258 kN (58000 lbf).
Continuous Rating: 2680 kW (3600 hp) giving a tractive effort of 89 kN (20000 lbf) at 67 m.p.h.
Maximum Rail Power: 4400 kW (5900 hp) at 38 m.p.h.

Brake Force: 40 t.	**Length over Buffers:** 17.83 m.
Design Speed: 100 m.p.h.	**Weight:** 83 t–84 t.
Max. Speed: 100 (75*) m.p.h.	**RA:** 6.
ETH Index: 74 (66§).	**Wheel Diameter:** 1156 mm.
Train Brakes: Air & Vacuum.	**Electric Brake:** Rheostatic.

Class 86/6 have the ETH equipment isolated.

86401		E	E	EW	CE	
86602	a*	F	FL	FL	CE	
86603	a*	FL	FL	FL	CE	
86604	a*	FL	FL	FL	CE	
86605	a*	FL	FL	FL	CE	
86606	a*	FL	FL	FL	CE	
86607	a*	F	FL	FL	CE	
86608	a*	RD	FL	FL	CE	
86609	a*	F	FL	FL	CE	
86610	a*	F	FL	FL	CE	
86611	a*	FL	FL	FL	CE	Airey Neave
86612	a*	FL	P	FL	CE	Elizabeth Garrett Anderson
86613	a*	F	P	FL	CE	County of Lancashire
86614	a*	FL	P	FL	CE	Frank Hornby
86615	a*	F	P	FL	CE	Rotary International
86416		RX	E	EW	CE	
86417		RX	E	EW	CE	
86618	a*	FL	P	FL	CE	
86419		RX	E	EW	CE	

86620	a*	F	P	FL	CE	
86621	a*	F	P	FL	CE	London School of Economics
86622	a*	FL	P	FL	CE	
86623	a*	FL	P	FL	CE	
86424		RX	E	EW	CE	
86425		RX	E	EW	CE	Saint Mungo
86426		E	E	EW	CE	
86627	a*	F	P	FL	CE	The Industrial Society
86628	a*	FL	P	FL	CE	Aldaniti
86430		RX	E	EW	CE	Saint Edmund
86631	a*	F	P	FL	CE	
86632	a*	F	P	FL	CE	Brookside
86633	a*	FL	P	FL	CE	Wulfruna
86634	a*	F	P	FL	CE	
86635	a*	F	P	FL	CE	
86636	a*	F	P	FL	CE	
86637	a*	FL	P	FL	CE	
86638	a*	FL	P	FL	CE	
86639	a*	F	P	FL	CE	

CLASS 87   BR DESIGN   Bo–Bo

Built: 1973–75 by BREL at Crewe Works.
Traction Motors: GEC G412AZ frame mounted (87/0), G412BZ (87/1).
Max. Tractive Effort: 258 kN (58000 lbf).
Continuous Rating: 3730 kW (5000 hp) giving a tractive effort of 95 kN (21300 lbf) at 87 m.p.h. (Class 87/0), 3620 kW (4850 hp) giving a tractive effort of 96 kN (21600 lbf) at 84 m.p.h. (Class 87/1).
Maximum Rail Power: 5860 kW (7860 hp) at 50.8 m.p.h.

Brake Force: 40 t.	**Length over Buffers:** 17.83 m.
Design Speed: 110 m.p.h.	**Weight:** 83.5 t.
Max. Speed: 110 (100*) m.p.h.	**RA:** 6.
ETH Index: 95 (75§).	**Wheel Diameter:** 1150 mm.
Train Brakes: Air.	**Electric Brake:** Rheostatic.

Class 87/0. Standard Design. Tap Changer Control.

87001	I	P	VW	WN	Royal Scot
87002	I	P	VW	WN	Royal Sovereign
87003	I	P	VW	WN	Patriot
87004	V	P	VW	WN	Britannia
87005	I	P	VW	WN	City of London
87006	V	P	VW	WN	George Reynolds
87007	I	P	VW	WN	City of Manchester
87008	I	P	VW	WN	City of Liverpool
87009	V	P	VW	WN	
87010	I	P	VW	WN	King Arthur
87011	I	P	VW	WN	The Black Prince
87012	I	P	VW	WN	The Royal Bank of Scotland

87013	I	P	*VW*	WN	John O' Gaunt
87014	I	P	*VW*	WN	Knight of the Thistle
87015	I	P	*VW*	WN	Howard of Effingham
87016	V	P	*VW*	WN	Willesden Intercity Depot
87017	I	P	*VW*	WN	Iron Duke
87018	I	P	*VW*	WN	Lord Nelson
87019	I	P	*VW*	WN	Sir Winston Churchill
87020	I	P	*VW*	WN	North Briton
87021	I	P	*VW*	WN	Robert the Bruce
87022	I	P	*VW*	WN	Cock o' the North
87023	I	P	*VW*	WN	Velocity
87024	V	P	*VW*	WN	Lord of the Isles
87025	I	P	*VW*	WN	County of Cheshire
87026	I	P	*VW*	WN	Sir Richard Arkwright
87027	I	P	*VW*	WN	Wolf of Badenoch
87028	I	P	*VW*	WN	Lord President
87029	I	P	*VW*	WN	Earl Marischal
87030	I	P	*VW*	WN	Black Douglas
87031	I	P	*VW*	WN	Hal o' the Wynd
87032	I	P	*VW*	WN	Kenilworth
87033	I	P	*VW*	WN	Thane of Fife
87034	I	P	*VW*	WN	William Shakespeare
87035	I	P	*VW*	WN	Robert Burns

Class 87/1. Thyristor Control.

| 87101 | * | | E | *EW* | CE | STEPHENSON |

CLASS 89   BRUSH DESIGN   Co–Co

Built: 1987 by BREL at Crewe Works.
Traction Motors: Brush design frame mounted.
Max. Tractive Effort: 205 kN (46000 lbf).
Continuous Rating: 2390 kW (3200 hp) giving a tractive effort of 105 kN (23600 lbf) at 92 m.p.h.
Maximum Rail Power:
Brake Force: 40 t.
Design Speed: 125 m.p.h.
Max. Speed: 125 m.p.h.
ETH Index: 95.
Train Brakes: Air.
Couplings: Drop-head buckeye.

Length over Buffers: 18.80 m.
Weight: 104 t.
RA: 6.
Wheel Diameter: 1150 mm.
Electric Brake: Rheostatic.

| 89001 | | **GN** | SL | *GN* | BN |

EWS liveried Class 58 No. 58033 prepares to leave Bilsthorpe Colliery on 21st March 1997 with a trip to High Marnham Power Station. Bilsthorpe Colliery ceased mining less than a month later. **Nic Joynson**

Class 59 No. 59204 'Vale of Glamorgan' approaches Gascoigne Wood with a train of coal hoppers on 2nd June 1997. Both the locomotive and wagons are in National Power livery.

John G. Teasdale

Mainline freight liveried Class 60 No. 60011 passes Spetchley on 9th May 1997 with the 07.31 Silverdale–Llanwern coal service.

Bob Sweet

▲ Class 73 No. 73131, in EWS livery, runs light from Stewarts Lane to Folkestone prior to working the Venice Simplon Orient Express on 1st May 1997.
Chris Wilson

▼ Class 86 No. 86101 'Sir William A Stanier FRS' passes Heastone Lane with the 16.00 London Euston–Manchester Piccadilly on 4th June 1997. The loco carries Intercity livery.
Chris Wilson

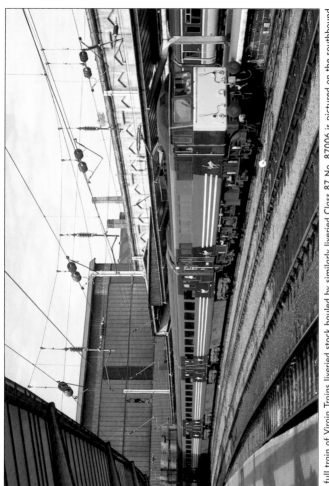

A full train of Virgin Trains liveried stock hauled by similarly liveried Class 87 No. 87006 is pictured on the southbound 'Royal Scot' service at Carlisle. The date is 20th September 1997.

Dave McAlone

▲ Great North Eastern Railway liveried Class 89 No. 89001 departs from London Kings Cross on 9th September 1997 with the 15.40 service to Bradford Foster Square.   **Hugh Ballantyne**

▼ The 17.35 Leeds–London Kings Cross is pictured near Sandal on 31st May 1997 with Class 91 No. 91014 providing the power at the rear. **John G. Teasdale**

Rail express systems liveried Class 90 No. 90016 arrives at its destination, London Kings Cross, with the 14.03 service from Low Fell on 6th September 1996.

Russell Ayre

Eurostar (UK) locomotive livery is carried by all Class 92s. One of them, 92003 'Beethoven' is pictured with the 13.45 Dollands Moor–Wembley at Kemsing on 1st May 1997.

Rodney Lissenden

CLASS 90   GEC DESIGN   Bo–Bo

Built: 1987–90 by BREL at Crewe Works. Thyristor control.
Traction Motors: GEC G412CY separately excited frame mounted.
Max. Tractive Effort: 258 kN (58000 lbf).
Continuous Rating: 3730 kW (5000 hp) giving a tractive effort of 95 kN (21300 lbf) at 87 m.p.h.
Maximum Rail Power: 5860 kW (7860 hp) at 68.3 m.p.h.

Brake Force: 40 t.	**Length over Buffers:** 18.80 m.
Design Speed: 110 m.p.h.	**Weight:** 84.5 t.
Max. Speed: 110 (75*) m.p.h.	**RA:** 7.
ETH Index: 95.	**Wheel Diameter:** 1156 mm.
Train Brakes: Air.	**Electric Brake:** Rheostatic.

Couplings: Drop-head buckeye (removed on Class 90/1).

Non-standard Liveries:
90128 is in SNCB/NMBS (Belgian Railways) electric loco livery.
90129 is in DB (German Federal Railways) 'neurot' (new red) livery.
90130 is in SNCF (French Railways) 'Sybic' livery.
90136 is **RD**, but with full yellow ends and roof and red 'Railfreight Distribution' lettering.

Class 90/0. As Built.

90001	I	P	*VW*	WN	BBC Midlands Today
90002	V	P	*VW*	WN	Mission:Impossible
90003	I	P	*VW*	WN	THE HERALD
90004	V	P	*VW*	WN	
90005	I	P	*VW*	WN	Financial Times
90006	I	P	*VW*	WN	High Sheriff
90007	I	P	*VW*	WN	Lord Stamp
90008	I	P	*VW*	WN	The Birmingham Royal Ballet
90009	I	P	*VW*	WN	The Economist
90010	I	P	*VW*	WN	275 Railway Squadron (Volunteers)
90011	I	P	*VW*	WN	The Chartered Institute of Transport
90012	V	P	*VW*	WN	British Transport Police
90013	I	P	*VW*	WN	The Law Society
90014	V	P	*VW*	WN	
90015	V	P	*VW*	WN	
90016	RX	E	*EW*	CE	
90017	RX	E	*EW*	CE	Rail express systems Quality Assured
90018	RX	E	*EW*	CE	
90019	RX	E	*EW*	CE	Penny Black
90020	E	E	*EW*	CE	Sir Michael Heron
90021	RD	E	*EW*	CE	
90022	RD	E	*EW*	CE	Freightconnection
90023	RD	E	*EW*	CE	
90024	RD	E	*EW*	CE	

Class 90/1. ETH equipment isolated.

90125	*	RD	E	EW	CE
90126	*	RD	E	EW	CE	Crewe International
						Electric Maintenance Depot
90127	*	F	E	EW	CE	Allerton T&RS Depot
						Quality Approved
90128	*	0	E	EW	CE	Vrachtverbinding
90129	*	0	E	EW	CE	Frachtverbindungen
90130	*	0	E	EW	CE	Fretconnection
90131	*	RD	E	EW	CE	Intercontainer
90132	*	RD	E	EW	CE	Cerestar
90133	*	RD	E	EW	CE	
90134	*	RD	E	EW	CE	
90135	*	RD	E	EW	CE	Crewe Basford Hall
90136	*	0	E	EW	CE	
90137	*	F	E	EW	CE	
90138	*	RD	E	EW	CE	
90139	*	F	E	EW	CE	
90140	*	F	E	EW	CE	
90141	*	FL	P	FL	CE	
90142	*	FL	P	FL	CE	
90143	*	FL	P	FL	CE	Freightliner Coatbridge
90144	*	FL	P	FL	CE	
90145	*	FL	P	FL	CE	
90146	*	FL	P	FL	CE	
90147	*	FL	P	FL	CE	
90148	*	FL	P	FL	CE	
90149	*	FL	P	FL	CE	
90150	*	FL	P	FL	CE	

CLASS 91   GEC DESIGN   Bo–Bo

Built: 1988–91 by BREL at Crewe Works. Thyristor control.
Traction Motors: GEC G426AZ.
Continuous Rating: 4540 kW (6090 hp).
Maximum Rail Power: 4700 kW (6300 hp).
Brake Force: 45 t.   **Length over Buffers:** 19.40 m.
Design Speed: 140 m.p.h.   **Weight:** 84 t.
Max. Speed: 140 m.p.h.   **RA:** 7.
ETH Index: 95.   **Wheel Diameter:** 1000 mm.
Train Brakes: Air.   **Electric Brake:** Rheostatic.
Couplings: Drop-head buckeye.

91001	GN	F	GN	BN
91002	GN	F	GN	BN
91003	GN	F	GN	BN
91004	GN	F	GN	BN
91005	GN ·	F	GN	BN

91006	**GN**	F	*GN*	BN	
91007	**GN**	F	*GN*	BN	
91008	**GN**	F	*GN*	BN	
91009	**GN**	F	*GN*	BN	The Samaritans
91010	**GN**	F	*GN*	BN	
91011	**GN**	F	*GN*	BN	
91012	**GN**	F	*GN*	BN	
91013	**GN**	F	*GN*	BN	
91014	**GN**	F	*GN*	BN	
91015	**GN**	F	*GN*	BN	
91016	**GN**	F	*GN*	BN	
91017	**GN**	F	*GN*	BN	
91018	**GN**	F	*GN*	BN	
91019	**GN**	F	*GN*	BN	
91020	**GN**	F	*GN*	BN	
91021	**GN**	F	*GN*	BN	
91022	**GN**	F	*GN*	BN	
91023	**GN**	F	*GN*	BN	
91024	**GN**	F	*GN*	BN	
91025	**GN**	F	*GN*	BN	
91026	**GN**	F	*GN*	BN	
91027	**GN**	F	*GN*	BN	
91028	**GN**	F	*GN*	BN	
91029	**GN**	F	*GN*	BN	
91030	**GN**	F	*GN*	BN	
91031	**GN**	F	*GN*	BN	

CLASS 92   BRUSH DESIGN   Co–Co

Built: 1993–5 by Brush Traction at Loughborough. Thyristor control.
Supply System: 25 kV a.c. from overhead equipment and 750 V d.c. third rail.
Electrical equipment: ABB Transportation, Zürich, Switzerland.
Traction Motors: Brush design.
Max. Tractive Effort: 400 kN (90 000 lbf).
Continuous Rating at Motor Shaft: 5040 kW (6760 hp).
Maximum Rail Power (25 kV a.c.): 5000 kW (6700 hp).
Maximum Rail Power (750 V d.c.): 4000 kW (5360 hp).
Brake Force: t.   **Length over Buffers:** 21.34 m.
Design Speed: 140 km/h (87½ m.p.h.).   **Weight:** 126 t.
Max. Speed: 140 km/h (87½ m.p.h.).   **RA:** 8.
ETH Index: 108.   **Wheel Diameter:** 1160 mm.
Train Brakes: Air.
Electric Brake: Rheostatic & regenerative.
Multiple Working: Time division multiplex system.
Communication Equipment: Driver–guard telephone.
Cab Signalling: Fitted with TVM430 cab signalling for Channel Tunnel.

92001	**EP**	E	*EW*	CE	Victor Hugo
92002	**EP**	E	*EW*	CE	H G Wells
92003	**EP**	E	*EW*	CE	Beethoven
92004	**EP**	E	*EW*	CE	Jane Austen
92005	**EP**	E	*EW*	CE	Mozart
92006	**EP**	CF	*EW*	CE	Louis Armand
92007	**EP**	E	*EW*	CE	Schubert
92008	**EP**	E	*EW*	CE	Jules Verne
92009	**EP**	E	*EW*	CE	Elgar
92010	**EP**	CF	*EW*	CE	Molière
92011	**EP**	E	*EW*	CE	Handel
92012	**EP**	E	*EW*	CE	Thomas Hardy
92013	**EP**	E	*EW*	CE	Puccini
92014	**EP**	CF	*EW*	CE	Emile Zola
92015	**EP**	E	*EW*	CE	D H Lawrence
92016	**EP**	E	*EW*	CE	Brahms
92017	**EP**	E	*EW*	CE	Shakespeare
92018	**EP**	CF	*EW*	CE	Stendhal
92019	**EP**	E	*EW*	CE	Wagner
92020	**EP**	LC	*EW*	CE	Milton
92021	**EP**	LC	*EW*	CE	Purcell
92022	**EP**	E	*EW*	CE	Charles Dickens
92023	**EP**	CF	*EW*	CE	Ravel
92024	**EP**	E	*EW*	CE	J S Bach
92025	**EP**	E	*EW*	CE	Oscar Wilde
92026	**EP**	E	*EW*	CE	Britten
92027	**EP**	E	*EW*	CE	George Eliot
92028	**EP**	CF	*EW*	CE	Saint Saëns
92029	**EP**	E	*EW*	CE	Dante
92030	**EP**	E	*EW*	CE	Ashford
92031	**EP**	E	*EW*	CE	
92032	**EP**	LC	*EW*	CE	César Franck
92033	**EP**	CF	*EW*	CE	Berlioz
92034	**EP**	E	*EW*	CE	Kipling
92035	**EP**	E	*EW*	CE	Mendelssohn
92036	**EP**	E	*EW*	CE	Bertolt Brecht
92037	**EP**	E	*EW*	CE	Sullivan
92038	**EP**	CF	*EW*	CE	Voltaire
92039	**EP**	E	*EW*	CE	Johann Strauss
92040	**EP**	LC	*EW*	CE	Goethe
92041	**EP**	E	*EW*	CE	Vaughan Williams
92042	**EP**	E	*EW*	CE	Honegger
92043	**EP**	CF	*EW*	CE	Debussy
92044	**EP**	LC	*EW*	CE	Couperin
92045	**EP**	LC	*EW*	CE	Chaucer
92046	**EP**	LC	*EW*	CE	Sweelinck

3. LOCOMOTIVES AWAITING DISPOSAL

03079		IL	Sandown
03179	**N**	IL	Ryde T&RSMD
08419		E	ADtranz Crewe Works
08473		E	Leicester LIP
08515		E	Gateshead WRD
08609		E	Willesden TMD
08618		E	Gateshead WRD
08634		E	Stratford TMD
08666		E	Allerton TMD
08673	**IO**	E	Allerton TMD
08677		E	Willesden TMD
08733		E	Motherwell TMD
08755		E	Millerhill Wagon Wks
08793	**0**	E	RFS(E) Ltd., Doncaster
08829		E	Toton TMD
08855		E	RFS(E) Ltd., Doncaster
08880		E	Allerton TMD
08895		E	Margam LIP
08898		E	RFS(E) Ltd., Doncaster
20073		E	Bescot Yard
20113	**0**	D	Brush, Loughborough
20119		E	Toton TMD
20154		E	Toton TMD
20175	**0**	D	Brush, Loughborough
20177		E	Toton TMD
25083		E	Crewe Brook Sidings
31105	**FT**	E	Bescot Yard
31112	**CT**	E	Bescot Yard
31168		E	Bescot Yard
31180	**FR**	E	Toton Yard
31184	**FO**	E	Toton Yard
31196	**C**	E	Stratford TMD
31209	**F**	E	Toton Yard
31217	**F**	E	Toton TMD
31282	**FR**	E	Bescot Yard
31283	**0**	E	Stratford TMD
31286		E	Bescot Yard
31289		E	Bescot Yard
31290	**C**	E	Toton Yard
31296	**F**	E	Crewe Brook Sidings
31299	**FO**	E	Stratford TMD
31320		E	Stratford TMD
31402		E	Bescot Yard
31403		E	Toton Yard
31428		E	Bescot Yard
31442		E	Crewe Brook Sidings
31460		E	Bescot Yard
31547	**C**	E	Toton Yard
31553	**C**	E	Toton Yard
31569	**C**	E	Toton Yard
33038		E	Stratford TMD
33205	**F**	E	Hither Green TMD
37252	**F**	E	Doncaster TMD
45015		E	Toton TMD
47096		E	Tinsley TMD
47102		E	Tinsley TMD
47190	**F**	E	Tinsley TMD
47214	**F**	E	Tinsley TMD
47249	**FR**	E	Tinsley TMD
47318	**FO**	E	Bescot Yard
47321	**F**	E	Tinsley TMD
47325	**FO**	E	Tinsley TMD
47515	**M**	E	Crewe Coal Siding
47707	**RX**	E	Crewe Basford Hall Yd
47714	**RX**	E	Crewe Basford Hall Yd
56009	**F**	E	Brush, Loughborough
56013	**F**	E	Toton TMD
56023	**F**	E	Toton TMD
56028	**F**	E	Margam WRD
56030	**F**	E	Margam WRD
56122	**F**	E	Toton TMD
97653	**0**	E	Reading T&RSMD

Non-Standard Liveries:
08793 is in London & North Eastern Railway apple green.
20113 & 20175 are RFS grey with blue and yellow bodyside stripes and carry numbers 2003 & 2007 respectively.
31283 is BR blue with large numbers.
97653 is Departmental yellow.

4. TOPS POOL CODES

A list of codes as used by TOPS is shown here for reference purposes:

SERCO

CDJD   Derby Etches Park Class 08

EWS (FORMERLY RAILFREIGHT DISTRIBUTION)

DAAN   Allerton Class 08
DADC   Crewe Electric Class 92 (Dollands Moor–Wembley)
DAEC   Crewe Electric Class 92 (Non-Operational)
DAET   Tinsley Class 47
DAIC   Crewe Electric Class 92 (EPS Testing)
DAMC   Crewe Electric Class 87/1 & 90
DASY   Tinsley Class 08/09 (Saltley)
DATI   Tinsley Class 08
DAWE   Allerton Class 08 (Wembley/Dagenham)
DAXT   Locomotives Awaiting Repair
DAYX   Stored Locomotives

FREIGHTLINER

DFLC   Crewe Electric Class 90/1
DFLM   Crewe Diesel Class 47 (Multiple Working Fitted)
DFLS   Crewe Diesel Class 08
DFLT   Crewe Diesel Class 47
DFNC   Crewe Electric Class 86/6
DFYX   Stored Locomotives
DHLT   Crewe Diesel Class 47 (Holding Pool)

EWS (FORMERLY MAINLINE)

ENAN   Toton Class 60
ENBN   Toton Class 58
ENSN   Toton Class 08/09 (Toton/Peterborough)
ENTN   Toton Class 37
ENXX   Stored Locomotives
ENZX   Locomotives For Withdrawal
EWDB   Eastleigh Class 37
EWDS   Eastleigh Class 33
EWEB   Hither Green Class 73
EWEH   Eastleigh Class 08
EWHG   Hither Green Class 37
EWOC   Old Oak Common Class 08/09
EWRB   Hither Green Class 73 (Restricted Use)
EWSF   Stratford Class 08/09
EWSU   Selhurst Class 08/09
EWSX   Stored/Reserve Shunters

EWS (FORMERLY LOADHAUL)

FDBI   Immingham Class 56

FDCI   Immingham Class 37
FDSD   Doncaster Class 08/09
FDSI   Immingham Class 08
FDSK   Knottingley Class 08/09
FDSX   Stored Shunters
FDYX   Stored Locomotives
FMSY   Thornaby Class 08/09

EUROSTAR (UK)

GPSN   Old Oak Common Class 73 (North Pole)
GPSS   Old Oak Common Class 08 (North Pole)
GPSV   Old Oak Common Class 37/6

TRAIN OPERATING COMPANIES

HASS   ScotRail Railways – Inverness Class 08
HBSH   Great North Eastern Railway – Bounds Green/Craigentinny Class 08
HEBD   Merseyrail Electrics – Birkenhead North Class 73
HFSL   Virgin West Coast – Longsight Class 08
HFSN   Virgin West Coast – Willesden Class 08
HGSS   Central Trains – Tyseley Class 08
HISE   Midland Mainline – Derby Etches Park Class 08
HISL   Midland Mainline – Neville Hill Class 08
HJSE   Great Western Trains – Landore Class 08
HJSL   Great Western Trains – Laira Class 08
HJXX   Great Western Trains – Old Oak Common/St Phillips Marsh Class 08
HLSV   Cardiff Railway Co. – Cardiff Canton Class 08
HSSN   Anglia Railways – Norwich Crown Point Class 08
HWSU   Connex South Central – Selhurst Class 09
HYSB   South Western Trains – Bournemouth Class 73/1
IANA   Anglia Railways – Norwich Crown Point Class 86/2
ICCA   Virgin Cross Country – Longsight Class 86/2
ICCP   Virgin Cross Country – Laira Class 43
ICCS   Virgin Cross Country – Edinburgh Craigentinny Class 43
IECA   Great North Eastern Railway – Bounds Green Class 91
IECB   Great North Eastern Railway – Bounds Green Class 89
IECP   Great North Eastern Railway – Craigentinny/Neville Hill Class 43
ILRA   Virgin Cross Country – Crewe Diesel Class 47/8
ILRB   Virgin Cross Country – Crewe Diesel Class 47/8 (Spot Hire)
IMLP   Midland Mainline – Neville Hill Class 43
IVGA   Gatwick Express – Stewarts Lane Class 73
IWCA   Virgin West Coast – Willesden Class 87/90
IWCP   Virgin West Coast – Manchester Longsight Class 43
IWLA   Great Western Trains – Laira Class 47
IWLX   Great Western Trains – Laira Class 47 (Reserve)
IWPA   Virgin West Coast – Willesden Class 86
IWRP   Great Western Trains – Laira/St Phillips Marsh Class 43

EWS (FORMERLY TRANSRAIL)

LBBS   Bescot Class 08/09
LCWX   Strategic Reserve Locomotives
LCXX   Stored Locomotives

LGBM   Motherwell Class 37
LGHM   Motherwell Class 37/4 (West Highland)
LGML   Motherwell Class 08/09
LNCF   Cardiff Canton Class 08/09
LNCK   Cardiff Canton Class 37 (Wales)
LNLK   Cardiff Canton Class 37 (St Blazey)
LNSK   Cardiff Canton Class 37 (Sandite Fitted)
LNWK   Cardiff Canton Class 08 (Allied Steel & Wire)
LWCW   Immingham Class 47
LWMC   Crewe Diesel Class 37/4 (North West Passenger)
LWNW   Bescot Class 31
LWSP   Crewe Diesel Class 08

HERITAGE LOCOMOTIVES

MBDL   Diesel Locomotives

EWS (FORMERLY RES)

PXLB   Crewe Diesel Class 47 (Extended Range)
PXLC   Crewe Diesel Class 47
PXLD   Motherwell Class 47
PXLE   Crewe Electric Class 86/90
PXLK   Crewe Diesel Class 47/9
PXLP   Crewe Diesel Class 47 (VIP Fleet)
PXLS   Crewe Diesel/Heaton/Old Oak Common Class 08/09
PXXA   Stored Locomotives

FORWARD TRUST RAIL

SAXL   Locomotives Off Lease

PORTERBROOK LEASING COMPANY

SBXL   Locomotives Off Lease

ANGEL TRAIN CONTRACTS

SCXL   Locomotives Off Lease

FRAGONSET RAILWAYS

SDFR   Locomotives For Hire

OTHER OPERATORS

XHSD   Direct Rail Services
XHSS   Direct Rail Services Stored Locomotives
XYPA   ARC Class 59/1
XYPD   Hunslet-Barclay Class 20/9
XYPN   National Power Class 59/2
XYPO   Foster-Yeoman Class 59/0
XYPS   Hunslet-Barclay Stored Locomotives

It will be possible, in most cases, for you to identify which pool a loco belongs to by its class and the owner, operation and depot codes shown. There are, however, a few exceptions to the rule and these are listed below.

Class 08 & 09

E owned locos whose operation code is *EW* and depot code is AN are DAAN except for 08389/93/482/694/825/44/72/913 which are DAWE.

E owned locos whose operation code is *EW* and depot code is CD are LWSP except for 08701/802/73/921, 09012 which are PXLS.

E owned locos whose operation code is *EW* and depot code is CF are LNCF except for 08651/819/900/32/42/54/5/93 which are LNWK.

E owned locos whose operation code is *EW* and depot code is TI are DASY except for 08879 which is DATI.

Class 37

E owned locos whose operation code is *EW* and depot code is CF are LNCK except for 37521/668–74/96 which are LNLK and 37178/97/229/30/54/63/75 which are LNSK.

E owned locos whose operation code is *EW* and depot code is ML are LGBM except for 37401/4/5/6/9/10/3/24/5/8/30 which are LGHM.

Class 47

E owned locos whose operation code is *EW* and depot code is CD are PXLB except for 47467/501/65/72/5/84/96/624/7/34/5/40 which are PXLC, 47971/6 which are PXLK and 47798/9 which are PXLP.

FL and P owned locos whose operation code is *FL* and depot code is CD are DFLT except 47114/50/2/7/204/5/9/34/58/79/87/9/90/2/303/9/23/30/7/58/61/70 which are DFLM.

P owned locos whose operation code is *GW* and depot code is LA are IWLA except for 47813/32 which are IWLX.

P owned locos whose operation code is *VX* and depot code is CD are ILRA except for 47825 which is ILRB.

Class 73

E owned locos whose operation code is *EW* and depot code is HG are EWEB except for 73128/31/2/9/40/1 which are EWRB.

Class 92

All locos are DADC except for 92005/8/14/7/8/21/3/5/7/36/40/3–6 which are DAEC and 92032 which is DAIC.

5. EUROTUNNEL LOCOMOTIVES

CLASS 0   MaK   Bo–Bo

These general purpose diesel locomotives are the same basic design as the Netherlands Railways 6400 Class.
Built: 1992–3 by Krupp-MaK/ABB at Kiel, Germany. (Type DE1004)
Engine: MaK 940 kW (1280 hp) at 1800 r.p.m.
Traction Motor: Four ABB three-phase traction motors.
Max. Tractive Effort: 305 kN.
Continuous Tractive Effort: 140 kN at 20 m.p.h.
Power at Rail: 750 kW.
Brake Force: 120 kN.   **Length over Couplers:** 16.50 m.
Weight: 84 tonnes.   **Wheel Diameter:** 1000 mm.
Max. Speed: 120 (75 km/h).   **Train Brakes:** Air.
Communication Equipment: Cab to shore radio.
Couplings: High and Low level Sharfenberg plus UIC screw.
Cab Signalling: TVM 430.
Livery: Standard NS grey and yellow (Netherlands Railways).

0001	ET	*ET*	CU		0004	ET	*ET*	CU
0002	ET	*ET*	CU		0005	ET	*ET*	CU
0003	ET	*ET*	CU					

CLASS 0   B

Schöma rebuilds of Hunslet 900 mm gauge locos.
Built: 1989–90. Rebuilt 1993–4.
Engine: Deutz 270 kW (200 hp).
Transmission: Mechanical.
Max. Speed: 30 m.p.h. (120 km/h when gears disengaged).
Train Brakes: Air.

0031	ET	*ET*	CU	FRANCES	0037	ET	*ET*	CU	LYDIE
0032	ET	*ET*	CU	ELIZABETH	0038	ET	*ET*	CU	JENNIE
0033	ET	*ET*	CU	SILKE	0039	ET	*ET*	CU	DIGITA
0034	ET	*ET*	CU	AMANDA	0040	ET	*ET*	CU	JILL
0035	ET	*ET*	CU	MARY	0041	ET	*ET*	CU	KIM
0036	ET	*ET*	CU	LAWRENCE	0042	ET	*ET*	CU	NICOLLE

CLASS 0   BRUSH EUROSHUTTLE   Bo–Bo–Bo

A.C. electric locomtives which are used on the Eurotunnel shuttle trains between Cheriton and Coquelles.
Built: 1992–4 by Brush/ABB at Loughborough.
Supply System: 25 kV a.c. from overhead equipment.
Traction Motors: 6 x ABB 6PH.
Control System: GTO thyristor.

Max. Tractive Effort: 400 kN (90 000 lbf).
Continuous Rating: 5760 kW (7725 hp) giving a tractive effort of 310 kN at 65 km/h.

Brake Force: 50 t.	**Length over Buffers:** 22.00 m.
Design Speed: 175 km/h (110 m.p.h.).	**Weight:** 132 t.
Max. Speed: 160 km/h (100 m.p.h.).	**RA:** Channel Tunnel only.
ETH Index:	**Wheel Diameter:** 1090 mm.
Train Brakes: Air	**Electric Brake:** Regenerative.

Multiple Working: Time division multiplex system. RC232 data Bus.
Couplings: High level Sharfenberg plus UIC screw links.
Communication Equipment: Cab to shore radio and in-train system.
Cab Signalling: TVM 430.
Livery: Two-tone grey and white with green, blue bands.

9001	ET	*ET*	CU	LESLEY GARRETT
9002	ET	*ET*	CU	STUART BURROWS
9003	ET	*ET*	CU	BENJAMIN LUXON
9004	ET	*ET*	CU	VICTORIA DE LOS ANGELES
9005	ET	*ET*	CU	JESSYE NORMAN
9006	ET	*ET*	CU	REGINE CRESPIN
9007	ET	*ET*	CU	DAME JOAN SUTHERLAND
9008	ET	*ET*	CU	ELISABETH SODERSTROM
9009	ET	*ET*	CU	FRANCOIS POLLET
9010	ET	*ET*	CU	JEAN-PHILIPPE COURTIS
9011	ET	*ET*	CU	JOSE VAN DAM
9012	ET	*ET*	CU	LUCIANO PAVAROTTI
9013	ET	*ET*	CU	MARIA CALLAS
9014	ET	*ET*	CU	LUCIA POPP
9015	ET	*ET*	CU	LÖTCHBERG
9016	ET	*ET*	CU	WILLARD WHITE
9017	ET	*ET*	CU	JOSE CARRERAS
9018	ET	*ET*	CU	WILHELMINIA FERNANDEZ
9019	ET	*ET*	CU	MARIA EWING
9020	ET	*ET*	CU	NICOLAI GHIAUROV
9021	ET	*ET*	CU	TERESA BERGANZA
9022	ET	*ET*	CU	DAME JANET BAKER
9023	ET	*ET*	CU	DAME ELISABETH LEGGE-SCHWARZKOPF
9024	ET	*ET*	CU	GOTTHARD
9025	ET	*ET*	CU	JUNGFRAUJOCH
9026	ET	*ET*	CU	FURKATUNNEL
9027	ET	*ET*	CU	BARBARA HENDRICKS
9028	ET	*ET*	CU	DAME KIRI TE KANAWA
9029	ET	*ET*	CU	THOMAS ALLEN
9031	ET	*ET*	CU	PLACIDO DOMINGO
9032	ET	*ET*	CU	RENATA TIBALDI
9033	ET	*ET*	CU	MONSERRAT CABALLE
9034	ET	*ET*	CU	MIRELLA FRENI
9035	ET	*ET*	CU	NICOLAI GEDDA
9036	ET	*ET*	CU	ALAIN FONDARY
9037	ET	*ET*	CU	GABRIEL BAQUIER
9038	ET	*ET*	CU	HILDEGARD BEHRENS

LIVERY CODES

Locomotives are BR blue unless otherwise indicated. The colour of the lower half of the bodyside is stated first. Minor variations to these liveries are ignored.

AC   ARC (yellow and grey with grey lettering and cast numberplates).

BR   Revised BR blue (blue with yellow cabs, grey roof, large numbers and full height BR logo).

BS   BR blue with red solebar stripe.

C   Civil Engineers (grey and yellow with black cab doors and window surrounds).

CS   Central Services (grey and red).

CT   Civil Engineers livery with Transrail lettering and markings (large white 'T' on a blue circle with a red outline underlined with red stripes).

D   Departmental (plain grey with black cab doors and window surrounds).

DR   Direct Rail Services (dark blue with light blue roof and green lettering).

E   English, Welsh & Scottish Railway (maroon with large maroon EW&S or EWS lettering and number on a broad gold band between cabs).

EP   European Passenger Services locomotive livery (as **F** with dark blue roof and cast Channel Tunnel logo).

F   New Railfreight (two-tone grey with black cab doors and window surrounds. Some locos still retain Trainload Coal, Construction, Metals, Petroleum or Railfreight Distribution markings).

FG   Fragonset Railways (black with a silver roof and a broad red stripe between the cabs).

FH   New Railfreight Livery with Loadhaul lettering.

FL   Freightliner (as **F** with black Freightliner lettering and red markings (diagonal stripes behind right hand cab door)).

FM   New Railfreight Livery with Mainline markings.

FO   Old Railfreight (grey sides, yellow cabs and full height BR logo).

FR   Old Railfreight Revised (as **FO** but with a red solebar stripe and a slightly smaller BR logo).

FT   New Railfreight Livery with Transrail lettering and markings (large white 'T' on a blue circle with a red outline underlined with red stripes).

FY   Foster-Yeoman (blue/silver/blue livery with white lettering and cast numberplates).

G   BR or GWR green.

GN   Great North Eastern Railway (dark blue with an orange bodyside stripe and gold or silver GNER lettering).

GW   Great Western Trains (green and ivory with Great Western Trains logo and lettering).

GX   Gatwick Express (white and dark grey with claret stripe and Gatwick Express lettering and motif).

HB   Hunslet-Barclay (two-tone grey with red solebars and black lettering).

I   InterCity (white and dark grey with red stripe and swallow motif).

IO   Old InterCity (light grey and dark grey with red stripe, yellow lower cab sides and BR logo).

LH   Loadhaul (black with orange cabsides and Loadhaul lettering).

M   Mainline (as **IO** but without the yellow lower cabsides and BR logo).

MD Merseyrail Departmental (dark grey and yellow with Merseyrail logo and lettering).

ML Mainline Freight (blue with silver body stripe and Mainline logo and lettering).

MM Midland Mainline (grey and green with three orange bodyside stripes and Midland Mainline logo and lettering).

N Network SouthEast (grey/white/red/white/blue/white).

NP National Power (grey/red/white and blue with white and red lettering and cast numberplates).

O Other livery (non-standard - refer to text).

PL Porterbrook Leasing (purple at one end, white at the other with small logo and lettering behind the left-hand cab doors. The livery represents an enlarged portion of the Porterbrook logo with the colours reversed on the other side).

R Parcels (post office red and dark grey).

RD New Railfreight Distribution (light grey and dark grey with dark blue roof, black Railfreight Distribution lettering and RfD markings (red diamonds on a yellow background)).

RR Regional Railways (grey/light blue/white/dark blue).

RX Rail express systems (post office red with Res blue & black markings)

ST Stagecoach (grey/orange/red/white/blue/white).

T Racal-BRT (two-tone grey with green markings).

V Virgin Trains (red with black cabs extending into bodysides, three white lower bodysides stripes and small Virgin logo behind cab doors on locos or red with black inner ends and large full height Virgin logo on Class 43 power cars).

W Waterman Railways (black with cream & red lining).

DIESEL & ELECTRIC LOCO REGISTER 3rd edition
by Alan Sugden.

Diesel & Electric Loco Register contains a complete listing of all non-steam locomotives ever possessed by British railways and its constituent companies. For each class of locomotive details of BR classification, builder, horsepower, transmission type and wheel arrangement are included, plus details of number, withdrawal date, brake type, train heating facilities and disposal for each individual locomotive. The book also has reserved space for the reader to enter his or her own information next to each record. 160 pages. A5 size. Thread Sewn. £7.95.

This title is available from the Platform 5 Mail Order Department. To order, please see centre pages of this book.

OWNER CODES

A	Angel Trains Contracts Ltd.
AC	ARC Limited
AD	ADtranz
AR	Anglia Railways
CA	Cardiff Railway Company
CF	Société Nationale des Chemins de Fer Francais (French Railways)
CT	Central Trains
D	Direct Rail Services
E	English Welsh & Scottish Railway Ltd.
ET	The Channel Tunnel Group Ltd.
F	Eversholt Holdings Forward Trust Rail Ltd.
FF	The Fifty Fund
FG	Fragonset Railways
FL	Freightliners Ltd.
FY	Foster-Yeomen Ltd.
GN	Great North Eastern Railway
GW	Great Western Trains
H	Hunslet-Barclay Ltd.
I	ICI Chemical & Polymers Ltd.
IL	Island Line (Isle of Wight)
LC	L&C Eurostar (UK) Ltd.
LN	London & North Western Railway Co. Ltd.
ME	Merseyrail Electrics
MH	Mid-Hants Railway plc
MM	Midland Mainline
MR	Midland & Northern Railroad Company Ltd.
NP	National Power
NT	9000 Trains
P	Porterbrook Leasing Co. Ltd.
RC	Railcare Ltd.
RF	RFS(E) Ltd.
SC	Connex South Central
SL	Sea Containers Rail Services Ltd.
SO	Serco (Furney Railtest)
SR	ScotRail
SW	South West Trains
VT	Virgin Trains
WT	Wessex Traincare Ltd.

OPERATION CODES

AD	ADtranz
AR	Anglia Railways
CA	Cardiff Railway Company
CT	Central Trains
DR	Direct Rail Services
ES	Eurostar (UK)
EW	English Welsh & Scottish Railway Ltd.
ET	The Channel Tunnel Group Ltd.
FL	Freightliners Ltd.
GN	Great North Eastern Railway
GW	Great Western Trains
GX	Gatwick Express
HB	Hunslet-Barclay Ltd.
IC	ICI Chemical & Polymers Ltd.
JF	Jarvis Facilities plc
MD	Mendip Rail
ME	Merseyrail Electrics
ML	Midland Mainline
NP	National Power
RC	Railcare Ltd.
SC	Connex South Central
SO	Serco (Furney Railtest)
SR	ScotRail
SS	Used normally on special or charter services
SW	South West Trains
VW	Virgin West Coast
VX	Virgin Cross Country
WT	Wessex Traincare Ltd.

DEPOT TYPE CODE

EMD	Electric Maintenance Depot
FP	Fuelling Point
LIP	Locomotive Inspection Point
SD	Servicing Depot
TMD	Traction Maintenance Depot
TMD (D)	Traction Maintenance Depot (Diesel)
TMD (E)	Traction Maintenance Depot (Electric)
TMD (HST)	HST Maintenance Depot
T&RSMD	Traction and Rolling Stock Maintenance Depot
WRD	Wagon Repair Depot

DEPOT & LOCATION CODES

AB	Aberdeen	KY	Knottingley TMD
AN	Allerton TMD (Liverpool)	LA	Laira T&RSMD (Plymouth)
AY	Ayr TMD	LE	Landore T&RSMD (Swansea)
BD	Birkenhead North T&RSMD	LG	Longsight TMD (E) (Manchester)
BH	Billingham (ICI) *	LO	Longsight TMD (D) (Manchester)
BK	Bristol Barton Hill T&RSMD	MD	Merehead (Foster Yeoman)
BL	Brush Traction, Loughborough *	MG	Margam
BM	Bournemouth T&RSMD	ML	Motherwell TMD
BN	Bounds Green T&RSMD (London)	NC	Norwich Crown Point T&RSMD
BS	Bescot TMD (Walsall)	NL	Neville Hill T&RSMD (Leeds)
BZ	St Blazey TMD	OC	Old Oak Common TMD (D)
CD	Crewe Diesel TMD	OO	Old Oak Common TMD (HST)
CE	Crewe International EMD	PB	Peterborough SD
CF	Cardiff Canton T&RSMD	PH	Perth
CL	Carlisle Upperby T&RSMD	PM	St. Phillips Marsh T&RSMD (Bristol)
CN	Carnforth *	PZ	Penzance
CO	Cranmore (East Somerset Rly)	RL	Ropley (Mid-Hants Railway)
CQ	Crewe (The Railway Age) *	SD	Sellafield (Direct Rail Services)
CU	Coquelles (Eurotunnel (France))	SF	Stratford TMD (London)
DR	Doncaster TMD	SL	Stewarts Lane T&RSMD (London)
DY	Derby Etches Park T&RSMD	SP	Springs Branch (Wigan)
EC	Craigentinny T&RSMD (Edinburgh)	SU	Selhurst T&RSMD (London)
EH	Eastleigh T&RSMD	SY	Saltley SD
EX	Exeter SD	TE	Thornaby TMD
FB	Ferrybridge (National Power)	TI	Tinsley TMD (Sheffield)
FH	Frodingham	TM	Tyseley Railway Museum
FW	Fort William	TO	Toton TMD (Nottinghamshire)
HG	Hither Green TMD	TS	Tyseley TMD (Birmingham)
HM	Healey Mills	TY	Tyne Yard
HT	Heaton T&RSMD (Newcastle)	WA	Warrington Arpley LIP
IM	Immingham TMD (Lincolnshire)	WH	Whatley (ARC Limited)
IS	Inverness T&RSMD	WN	Willesden TMD (London)
KR	Kidderminster (Severn Valley Rly)		

* Unofficial code

WORKS CODES

ZB	RFS (E) Ltd., Doncaster
ZC	ADtranz Crewe Works
ZD	ADtranz Derby Carriage Works
ZF	ADtranz Doncaster Works
ZG	Wessex Traincare Ltd., Eastleigh Works
ZH	Railcare Ltd., Springburn Works, Glasgow
ZI	ADtranz Ilford Works
ZK	Hunslet-Barclay Ltd., Kilmarnock Works
ZN	Railcare Ltd., Wolverton Works
ZT	ADtranz Trafford Park Wheel Works, Manchester (unofficial code)